T0353575

Maatian Ethics in a Communication Context

Maatian Ethics in a Communication Context explores the ethical principle of Maat: the guiding principle of harmony and order that permeated classical African political and civil life.

The book provides a rigorous, communication-focused account of the ethical wisdom ancient Africans cultivated and is evidenced in the form of recovered written texts, mythology, stelae, prescriptions for just speech, and the hieroglyphic system of writing itself. Moving beyond colonial stereotypes of ancient Africans, the book offers insight into the African value systems that positioned humans as inextricably embedded in nature, and communication theory that anchors good communication in careful listening habits as the foundational moral virtue. Expanding on the work of Maulana Karenga, Molefi Kete Asante and other groundbreaking scholars, the book presents a picture of civilizations with a shared lust for life, a spiritual connection to scientific speech, and the veneration of ancestors as deeply connected to the pursuit of wisdom.

Offering an examination of Maat from a specifically communication ethics perspective, this book will be of great interest to scholars and students of Communication Ethics, African philosophy, Rhetorical theory, Africana Studies and Ancient History.

A native of Puerto Rico, **Melba Vélez Ortiz** earned her PhD in Communication Ethics from the Institute of Communications Research at the University of Illinois at Urbana-Champaign in 2009. Her areas of research are communication ethics and global environmental communication. Her work examines the ways in which the long-term success of conservation efforts depends upon fundamental shifts in cultural values, in aesthetic and moral communication, and in shared understandings of how the individual fits into social and ecological communities. She has a passion for intellectual history and ancient approaches to ethics. Her current research examines the ecological and communicative dimensions of ancient Pan-African ethics and values. She is Past-Chair of the Ethics Division of the National Communication Association and her research has been recognized and awarded at the national level. In addition, Dr. Vélez Ortiz has researched and published in the area of Latin-American/Caribbean/Latina-o philosophy and intellectual history.

Routledge Focus on Communication Studies

A Relational Model of Public Discourse
The African Philosophy of Ubuntu
Leyla Tavernaro-Haidarian

Communicating Science and Technology through Online Video
Researching a New Media Phenomenon
Edited by Bienvenido León and Michael Bourk

Strategic Communication and Deformative Transparency
Persuasion in Politics, Propaganda, and Public Health
Isaac Nahon-Serfaty

Globalism and Gendering Cancer
Tracking the Trope of Oncogenic Women from the US to Kenya
Miriam O'Kane Mara

Maatian Ethics in a Communication Context
Melba Vélez Ortiz

Maatian Ethics in a Communication Context

Melba Vélez Ortiz

Routledge
Taylor & Francis Group

NEW YORK AND LONDON

First published 2020
by Routledge
52 Vanderbilt Avenue, New York, NY 10017

and by Routledge
2 Park Square, Milton Park, Abingdon, Oxon OX14 4RN

Routledge is an imprint of the Taylor & Francis Group, an informa business

British Library Cataloguing-in-Publication Data
A catalogue record for this book is available from the British Library

Library of Congress Cataloging-in-Publication Data
A catalog record has been requested for this book

ISBN: 978-0-367-34482-5 (hbk)
ISBN: 978-0-429-32611-0 (ebk)

Typeset in Times New Roman
by codeMantra

Reproduction of material from *The Literature of Ancient Egypt* (© William Kelly Simpson, 1972) by kind permission of Yale University Press

Reproduction of material from *Ancient Egyptian Literature, Volume I: The Old and Middle Kingdoms* (© Miriam Lichtheim, 2006) by kind permission of University of California Press

Contents

Preface   vii
    As Above, So Below, As Below, So Above vii
    An Undeniable Moral, Philosophical,
        and Intellectual Legacy x
    Notes on the Particularities of This Inquiry xvi
Acknowledgments   xviii

**1   The Evolution of Maat as the Collective Guiding
     Principle in Pre-Colonial African Civilizations**   1
    The Earthshattering End of the Last Ice Age 3
    A Genealogy of Morals 5
    Ecological Dimensions 8
    Conclusion: Maat as a Land Ethic 12

2   Ancient African Spirituality: Heaven on Earth   16
    The Communicative Dimensions of African
      Spirituality 18
    From Cosmology to Communication Ethics 23
    The Kemetic Work of the Soul 27
    Conclusion: Mythology as Allegory 29

3   Scientific Communication and the Divine   34
    Monologue as Internal Dialogue 37
    Maat as the Model of Maieutic Communication 38
    Direct Communication as Divine 43
    The School of Alexandria 45
    Conclusion: Morality as Technology 46

4 The Universal Moral Ideal of Maat   50
*Re-Enlightening the Enlightenment
    Transcendental Individual 52
Communication as Key to Survival 55
On Reasoned Pleasures 58
Maatian Symmetry and Balance as Visual Ethics 62
Conclusion: Unaltered by the Winds of Change 64*

5 Communicative Dimensions of Maat: Speech and Silence   67
*Speech as a Radically Creative Action 67
Maat as a Scientific Speech 71
Speaking to Posterity: On the Preeminence
    of Listening and Silence as Communicative Virtues 77*

6 Medu Netcher: A Picture Says a Thousand Words?   80
*Order 82
Harmony 84
Balance 85
Transcendental Truths 86*

7 A Lust for Life: Allegory and Poetry   94
*Maat in the Ancestors' Voice 95
Love 96
Family 98
Sorrow 99
Friends 102
Nature 104*

Afterword: Speaking to Posterity   111

Index   115

Preface

As Above, So Below, As Below, So Above

Figure I.1 Funerary papyrus showing Egyptian cosmology.

As noted by communication, language, and rhetoric scholar Clinton Crawford, ancient Egyptian society was steeped in many principles that guided action, including the principle of correspondence. The cultural context that gives rise to Maat (*m3ꜥ*) is one with deep-rooted cosmological attitudes. However, given the contemporary zeitgeist privileging of self-determination, we must not err in interpreting a cosmological, divinely ordered human experience as primitive. This conflation is not only the result of insidious cultural, ethic, and racial biases but also a failure to understand human existence as grounded in material, ecological, economic, and cultural realities.

The pursuit of harmony, balance, and righteousness can indeed be perceived as constricting of human aspirations, imagination, and will... Still, there is another, no less important, dimension of maintaining harmony and balance that speaks directly to our self-knowledge and to the ability the human species has to know how they fit into their environment. Take, for example, Charles Darwin's evolutionary theory and the common interpretation of the phrase "survival of the fittest". A superficial understanding of this phrase will place an erroneous emphasis on dominance, aggression, and force as a way to ensure the survival of a species or its individual members. The more accurate interpretation of this phrase will emphasize the "fit" aspects of how a species or an individual matches its environment. The fit we are really talking about here is the degree to which a species meshes, is well suited, or can adapt to its material conditions.

In this respect, cosmology also places a demand on adaptability. If we take cosmology "to cover analyses, theories and explanations of the phenomena of the universe" (Wright 4), and we also understand that from these very analyses, humans often take their cues on how to preserve, extend, and enhance their lives and their environment..., we will inevitably run into practices then that run counter to the goal of preserving, and enhancing human life on the planet. In this respect, concepts of freedom and cosmology both are predicated on the limitations of individual freedom. For, while an emphasis on maintaining peace and stability in a society calls for the diminution of individual freedoms for the purpose of maintaining a certain social order, the very idea of freedom is premised on the fact that said freedom must not infringe on another individual's freedom, which, of course, requires some curtailment of those very freedoms.

Having established cosmology as inherently placing boundaries and confines on human behavior, depending on what narrative the cosmology supports in terms of origin and attributes, we can now understand the connection between cosmology and correspondence. Egyptian society was largely invested in permanence and continuity in all areas of cultural, technological, spiritual, and intellectual achievement. As Djehuti, the Egyptian deity highlighting the intellectual and scientific dimension of our human experience (also known as Tehuti and Thoth, and later as Hermes by the Greeks) noted, "as above so below, as below so above". In order words, ancient Egyptian society was consumed with understanding, mimicking, and expressing the ultimate nature of reality.

The historical records bear that ancient Egyptian people were engrossed by celestial body movements, substances, and forces, and they devoted an astounding amount of time to refining their knowledge of astronomy (Clarke 63). Their interest in astronomy can be attributed to many factors. Among them are the simultaneous advances in the areas of agriculture and navigation (Relke and Ernest 64; Stieglitz 134).

Such practices, when successful, require careful attention to the movements of celestial bodies, to climatological conditions, and to the fluctuations of the sun. Another important reason the Egyptians dedicated themselves to the study of the sky is tied to cosmology. Representing its powerful encircled disk (halo) by the deity of Atum, son of Ptah, the primordial creative force, Atum is attributed with having ordered the world out of the chaos of the primordial waters. According to folklore, the sundisk, while an optical illusion, can predict weather conditions (Macherera et al. 58), and this predictive ability is precisely the kind of scientific endeavor in which humans are engaged to this very day.

At its core, we value scientific knowledge for its ability to identify patterns and predict phenomena. This is why we value scientific pursuits—because this allows humans to make inferences that can yield predictions. The ancient Egyptians were no different in this regard. Theirs was a society that revolved around the study of its surroundings. It was also a society that wielded tremendous economic and cultural influence in friend and foe alike. In this respect, the figure of the sundisk, with its predictive, magnetic attributes, became an important symbol of Egyptian values. Science, observation, and the making of inferences and predictions became the collective behavioral model to be promoted among its citizenry. Knowledge, in this view, can order the world out of chaos and provide the stability that comes with prediction as it allows one to prepare, to forecast, to adapt, to know, and to become a better fit for one's environment.

"As above so below" kept the human gaze fixed on the search for patters in the sky that could shed light on our own human place in the cosmos. There is an a-priori assumption here that we, too, are ruled by the forces and laws, much like those that regulate nature and the cosmos as a whole. The second part of Djehuti's principle "as below so above" reinforces this idea that the terrestrial and celestial landscapes mirror one another in their ultimate subordination to the forces of creation. Thus, the principle of correspondence establishes a cosmological and inexorable link between the human and its environment, between the heavens and the earth.

Interestingly, this will not be the last time in history when a popular ethical approach will be grounded in this principle of correspondence. In not so many words, Immanuel Kant will advance the same argument in his defense of the deontological approach. In the *Groundwork to the Metaphysics of Morals*, Kant will justify a human moral order on the order that is evident in the phenomenal world. Indeed, the cosmology the guided Egyptian society in ancient times is not so different from the enlightenment recognition that nature is governed by mechanistic, causal processes (laws) that in their functionality and arrangement ought to mirror the practice of morality.

Whether we choose to talk about this in terms of the transcendental deduction or the purposiveness of nature in Kant's *Critique of Pure Reason*, the fact remains that Kant discourages incursions into the nature and structure of the "noumenal" world, the part of our human experience that is unseen and to which only God is privy, in favor of a focus on the "phenomenal" world, the world that we experience, that we can study, the empirical world.

The ancient Egyptians saw their environment as a source of wisdom, of insight into the human condition, and as the key to their survival and spiritual and intellectual enlightenment. The quest for human permanence and continuity is a motif that appears time and time again in the pages that follow. It is an end-goal that permeated every aspect of ancient Egyptian society and for which many a contemporary of this civilization expressed great admiration, and it continues to fascinate and amaze to this day. Even the casual reader is likely to have some image or reaction when the subject of ancient Egypt is brought up. Regrettably, the egyptomania industry has had such success that many of us are inundated with imagery of tombs, gods, and even aliens when we think about ancient Egypt, but this is a fairly recent phenomenon (Bernal 24). The truth is that ancient Egypt has been heralded throughout history for far more than its technological, militaristic, and funerary achievements, as we will see in the handful of quotes discussed below.

An Undeniable Moral, Philosophical, and Intellectual Legacy

One can hardly think of a greater contributor to universal history than the Greek historian Diodorus Siculus. Best known for his *Bibliothēkē*, the surviving portions of his original 40 volumes on the library of history have been a go to source for insight into our ancestral history since Roman times. Diodorus is said to have travelled throughout Egypt for three years from 60 to 57 AC (Sperry 150), and his recorded observations on this society provide a unique and, by today's standards, unusual account of the achievements and merits of ancient Egyptian civilization in so far as he credits the Egyptian writing system for its novel and symbolic attributes:

> We must now speak about the Ethiopian writing which is called hieroglyphic among the Egyptians, in order that we may omit

nothing in our discussion of their antiquities. Now it is found that the forms of their letters take the shape of animals of every kind, and of the members of the human body, and of implements and especially carpenters' tools; for their writing does not express the intended concept by means of syllables joined one to another, but by means of the significance of the objects which have been copied and by its figurative meaning which has been impressed upon the memory by practice. For instance, they draw the picture of a hawk, a crocodile, a snake, and of the members of the human body—an eye, a hand, a face, and the like (97).

It is important to note that Diodorus is not describing the Egyptian system of writing as though it was inferior to the Greeks in any way. His description is objective and earnest. Perhaps we can understand the reason for this deference by reviewing the words of another principal figure in Greek thought: Plato. While many scholars read the *Phaedrus* looking for insights into the Greek concepts of love, persuasive speech, truth, and even the nature of dialogue, some of the most interesting revelations made in the *Phaedrus* have to do with its plain and effusive references to the intellectual dominance of Egyptian culture, even during Socrates' time. He says:

I can tell you something I've heard, from people before me; only they know the truth . . . Well, what I heard was that one of the ancient gods of Egypt was at Naucratis in that country, in Egypt, the god to whom the sacred bird they call the ibis belongs; the divinity's own name was Theuth. The story was that he was the first to discover number and calculation, and geometry and astronomy, as well as the games of draughts and dice and, to cap it all, letters (61–62).

The implications of this quote are significant. Here, we have one of the leading philosophers in the history of philosophy, plainly asserting that it is the Egyptians, not the Greeks or anybody else, who invented mathematics, astronomy, and writing. The statement is offered with confidence and without caveat. It is a startling contrast to the way in which the illustrious intellectual history of the ancient Africans is largely denied and concealed in the 21st century. Plato's bold assertion through the voice of Socrates lends an additional layer of gravitas as Socrates was famously Plato's mentor and teacher. The almost casual manner in which Plato writes this account into this dialogue gives the appearance of an uncontested nature to these statements. It seems that the Greeks did not vacillate on many occasions to recognize the Egyptians as intellectual leaders, not simply crafty builders. Take, for example, the words of Aristotle: "Egyptians are reputed to be the oldest of nations, but they have always had laws and a political system" (419). In *The Politics*, Aristotle gives credit where credit is due and presents an image of ancient Africa as an epicenter of thought and a model of government.

Aristotle takes us far away from the modern caricature of ancient African people as uncivilized and offers yet one more piece of evidence that today's narrative is filled with white supremacist falsehoods and that, in fact, according to the Greeks at least, it was Africans who civilized the rest of the world, not the other way around. One need not look further than Plato's own words as he continues his account of history by, once again, elevating ancient Egyptian wisdom without hesitation. He states,

> It is reported that Thamus expressed many views to Theuth about each science, both for and against; it would take a long time to go through them in, but when they came to subject of letters, Theuth said, "but this study, King Thamus, will make the Egyptians wiser and improve their memories; what I have discovered is an elixir of memory and wisdom" (61).

Here, we have another attribution to the Egyptians of the invention of writing as well as further praising of Theuth (Thoth), as he is presented as a wise man who is steeped in science and who can both critically analyze and speak eloquently on a highly intellectual topic. However effusive the praise to the Egyptians by the ancient Greeks, the contemporary archeological record and conventional wisdom credit the invention of writing to the cuneiform of the Sumerian people in ancient Mesopotamia. Specifically, the Kesh Temple Hymn, often called "Liturgy to Nintud", has been carbon dated to around 2600 BC (Barton 2), while Egypt's pyramid texts have been dated to 200 years later, between 2400 and 2300 BC (Forman and Quirke 1).

While, at first, this evidence may appear to contradict the Greek account, scholars have noted the peculiarities and possible ethnic connection between Sumerians and the Kush people of ancient Nubia. For instance, Sumerians called themselves the Black-headed people and are thought to have been one of the many Nilotic Kushite people who migrated to set up colonies in early Asia (Rashidi 216). Furthermore, scholars have looked at biblical accounts where Nimrod is specifically referred to as a son of Kush, who the Roman and Greeks called "Ethiopians" and who originally inhabited modern-day Sudan (Rashidi 216). Lastly, it is worth noting that, as its neighbor to the south, Egypt forged uneasy and, at times, hostile political ties with the Kush while acknowledging Kush as its own ancestral motherland (Levi 179).

So far, we have heard Greek intellectual giants like Diodorus Siculus, Plato, and Aristotle speak plainly about the intellectual might of the ancient Africans, through their descriptions of ancient Egyptian wisdom, but what of the art and science of philosophy itself? Ethics, after all, is a branch of philosophy, and, in order to locate Maat as either the first or one of the first human ethical views, it is necessary to inquire into the origin of philosophy itself.

As arguably the greatest of all the sophists, Isocrates fits the bill. In his dialogue the *Busiris* (written around 390 BCE), Isocrates speaks plainly about who is responsible for the creation of philosophy, and once again, we find the Egyptians being credited in this endeavor: "All men agree the Egyptians are the healthiest and most long of life among men; and then for the soul they introduced philosophy's training..." (115). In his presentation of the wisdom of his time, Isocrates speaks unequivocally, philosophy was invented by African people. That this is anywhere taboo in the 21st century is more a product of the biases of our time and the complicity of the academy than anything else.

Isocrates has much more to say on this topic. He confirms what we already knew about Greek thinkers of statute, like Pythagoras, learning and studying the ancient Egyptians. He states:

> If one were not determined to make haste, one might cite many admirable instances of the piety of the Egyptians, that piety which I am neither the first nor the only one to have observed; on the contrary, many contemporaries and predecessors have remarked on it, of which Pythagoras of Samos is one. On a visit to Egypt he became a student of the religion of the people, and was the first to bring Greeks all philosophy, and more conspicuously than others he seriously interested himself in sacrifices and ceremonial purity, since he believed that even if he should gain thereby no greater reward from the gods, among men, at any rate, his reputation would be greatly enhanced. And this indeed happened to him (119).

In this quote, Isocrates adds an important dimension to the ancient relationship between the classical Greek and Classical Egyptian eras: Greeks looked upon these ancient Africans as role models of conduct. One cannot get closer to ethics than the admission that a civilization is renowned for their virtuous character, especially when such praise comes from a civilization that we generally consider to be the cradle of civilization. Isocrates' words leave no doubt. It was the ancient Africans who civilized the rest of the world, and not just with their scientific and technological achievements but through their approach to character building, which is to say ethics. Perhaps, the father of history, Herodotus, summarizes best when he says, "Concerning Egypt itself I shall extend my remarks to a great length, because there is no country that possesses so many wonders, nor any that has such a number of works which defy description" (138). That we look back at the ancient Egyptians with a distorted view of a culture steeped in superstition and death is more a reflection of who we are than who they were. Diodorus reports on the uniqueness of the Egyptian Medu Netcher (hieroglyphs) not with condescension, but with interest.

The study of human communication ethics is not complete without a nuanced, accurate account of the foundational contributions of ancient African cultures. Thus, the work that follows is a conscious and deliberate attempt to add my voice to the many scholars inside and outside the academy laboring to produce a rigorous, fact-based, unapologetic corrective history that breaks through the conspiracy of silence that has kept ancient African intellectual history from assuming its proper place in the cannon of all traditional disciplines and figuring prominently in existing and emerging fields of knowledge.

All history is a current event, according to Dr. John Henrik Clark, whose work in African intellectual history is required reading for those interested in this topic and serves as a lighthouse to those who undertake the difficult and righteous work of documenting and promoting this hidden history. My objectives in this inquiry are as follows: (1) To highlight the ethical and moral zeitgeist of ancient Egypt, a cluster of cultures that, although well known to the public and scholars for the technological, political, and cultural achievements, remain obscure in terms of their ethical commitments; (2) to help cement Maat as a legitimate approach to ethics as practiced and idiosyncratic as the Greek approach to virtue, Kant's deontology, and Jon Stuart Mill's consequentialism; and (3) to accent the Maatian emphasis on listening practices in interpersonal communication. Such focus that stands in contrast to the Greco-Roman emphasis on good speaking skills as the foundation of excellent and ethical communication.

In the field of communication ethics, and multidisciplinary ethics at large, research and teaching on ancient ethics is largely focused on the ancient Greeks. Undeniably, Socrates, Plato, Aristotle, and Isocrates are towering figures in the world of ancient ethics, and their contributions have stood the test of time. Classical approaches, like the virtue approach, continue to resonate with renowned contemporary ethics scholars, such as Allister McIntyre, Rosalind Hursthouse, Phillipa Foot, and Martha Nussbaum. Contemporary communication theory, including communication ethics, is inextricably intertwined with the wisdom of the ancient world and in many ways remains a blueprint for how we theorize and practice human communication.

However, in maintaining a narrow focus on ancient Greek ethics, we are missing a large swath of the history of ethics and, in doing so, omitting a significant cultural contribution to the canon. Greek historians like Diodorus, and more famously, Herodotus, often footnoted their work with references to the influence of ancient Egyptian priests on Greek philosophy and mythology. So, why is ancient Egyptian wisdom (different from ingenuity), values, and ethics not taught in ethics courses? Why don't I teach this in my own communication ethics courses? Must ancient Egyptian ethics remain a footnote in the intellectual history of both ethics and communication ethics? This book aims at making Maatian Egyptian ethics conversant with the field of communication ethics beyond rhetorical or philosophical analysis. My research question is: What is a Maatian Egyptian Ethic, and what does it have to teach us about good communication?

It is a fact that some Greek philosophers, like Thales, studied in Africa (de Mooij 2013), and communication scholars such as Molefi Kete Asante have already acknowledged, traced, and celebrated the cross-pollination between Greek and Egyptian "sophia" or wisdom. Asante is one of the few rhetoric and communication scholars who has made strides to incorporate the ancient Egyptian traditions into emerging communication research and teaching. Other indirectly relevant but notable contributions include Silvia Codita's *Ancient Egyptian Rhetoric* (2012) and Carol Lipson's *Ancient Egyptian Rhetoric: It All Comes Down to Maat* (2004). The work done by these scholars advanced our understanding of the rhetorical systems of ancient Egypt and emphasized three major areas: Writing and literacy, rhetorical norms and practices, and possible connections between Egyptian and Greek rhetorical precepts. The present work has benefitted greatly from these and other excellent studies. My hope is to offer the reader a narrower, more specific focus on what Maat have to say about: (1) The purpose or ends of communication, (2) ways in which we ought to communicate (acceptable means for desired ends), (3) ways in which we ought not to communicate (akin to unethical communication practice), and (4) the values embedded within ancient Egyptian writing and speaking practices.

Notes on the Particularities of This Inquiry

In the interest of historical accuracy and for the purposes of helping to decolonize and advance an area of study that has largely been suppressed, I will engage in three different ethical practices for the duration of this inquiry:

1   I will henceforth refer to ancient Egypt by the chosen name of its ancient people: Kemet (KMT). In doing so, I will be rejecting the Greek name Aegyptus (Egypt) as a deference to the people of this great civilization.
2   I will not make reference in this book to Egyptian gods and goddesses, and instead will refer to them as divinities or neter(u). Through my research, I have come to understand that the field of Egyptology came of age during the 18th and 19th centuries, and was founded and led by white, European, affluent men who in no small measure interpreted Egyptian values through their own religious lens, thereby turning many of the archeological discoveries and saturating text translations with a religious structure that simply was not in existence in the classical period of ancient Egypt. Instead, I will make reference to deities, divinities, or aspects of the human experience as a translation for neter, which contemporary scholars, among them historian Ashar Kwesi, have begun to see as the more accurate terminology that reflects the existence of a robust spirituality characteristic of the period without the connotation of a formal religion.
3   Finally, for the same reasons, I will avoid the name "Egypt"; henceforth, I will also refer to the language of the people of KMT not as hieroglyphs (Greek term for sacred text) but as Medu Netcher (divine speech), the name of their own choosing.

Bibliography

Aristotle. *The Politics.* Translated by T. A. Sinclair. Penguin Classics, 1981.

Asante, Molefi K. "A Discourse on Black Studies: Liberating the Study of African People in the Western Academy." *Journal of Black Studies*, vol. 36, no. 5, 2006, pp. 646–662.

Asante Molefi K. and Ama Mazama. *Egypt vs. Greece in the America Academy.* African American Images, 2002.

Barton, George A. *Miscellaneous Babylonian Inscriptions.* vol. 1. Nabu Press, 2010.

Bernal, Martin. *Black Athena: The Afroasiatic Roots of Classical Civilization.* Rutgers University Press. 1987.

Clarke, Leonard W. "Greek Astronomy and Its Debt to the Babylonians." *The British Journal for the History of Science*, vol. 1, no 1, 1962, pp. 65–77.

Crawford, Clinton. "The African Origin of Nubian/Egypt Rhetoric and Its Application to Contemporary Classrooms." *African American Rhetorics: Making the Invisible Seen*, edited by Ron L. Jackson and Elaine B. Richardson. Southern Illinois University Press, 2004, pp. 111–135.

———. *Recasting Ancient Egypt in the African Context*. Africa World Press, 1996.

Diodorus Siculus. *Diodorus Siculus: Library of History*, Volume I, Books 1–2.34 (Loeb Classical Library No. 279). Harvard University Press, 1933.

Forman, Werner and Stephen Quirke. *Hieroglyphs and the Afterlife in Ancient Egypt*. University of Oklahoma Press, 1996.

Hermes Trismegistus. *The Divine Pymander: And Other Writings of Hermes Trismegistus*. Translated by John D. Chambers. Martino Fine Books, 2018.

Herodotus. *The Histories*. Translated by George Rawlinson. Everyman's Library, 1997.

Isocrates. *Isocrates: Evagoras. Helen. Busiris. Plataicus. Concerning the Team of Horses. Trapeziticus. Against Callimachus. Aegineticus. Against Lochites. Against Euthynus. (Loeb Classical Library)*. Translated by George Norlin. Harvard University Press, 1928.

Levi, Josef Ben. "The Intellectual Warfare of Dr. Jacob H. Carruthers and the Battle for Ancient Nubia as a Foundational Paradigm in Africana Studies: Thoughts and Reflections". *Journal of Pan African Studies*, vol. 5, no. 12, 2012, 178–195.

Macherera, Margaret et al. "Indigenous Environmental Indicators for Malaria: A District Study in Zimbabwe." *Acta Tropica*, vol. 175, 2017, pp. 50–59.

Plato. *Phaedrus (Penguin Classics)*. Translated by Christopher Rowe. Penguin Classics, 2005.

Rashidi, Runoko. "The Kushite Origins of Sumer and Elam." *Ufahamu: A Journal of African Studies*, vol. 12, no. 3, 198, pp. 215–233.

Relke, Joan and Allan Ernest. "Ancient Egyptian Astronomy: Ursa Major—Symbol of Rejuvenation." *Archeoastronomy*, vol. 17, 2002, pp. 64–80.

Stieglitz, Robert R. "Long-Distance Seafaring in the Ancient near East." *The Biblical Archeologist*, vol. 47, no. 3, 1984, pp. 134–142.

Wright, M. R. *Cosmology in Antiquity*. Routledge, 1995.

Acknowledgments

This work would not be possible without the encouragement of Professor Kaba Hiawatha Kamene, Dr. Maulana Karenga, Dr. Molefi K. Asante, Dr. Clyde Winters, and Asar Imhotep. I am indebted to these gentlemen for their wisdom and generosity. I would also like to take this opportunity to thank my superb editor Suzanne Richardson and my teachers and mentors, all of whom played a part in cultivating my voice: Michael A. Weinstein, Diane S. Rubenstein, Paul Dale, William McBride, Patricia Curd, Zine Magubane, Karen Whedbee, Mary Keener, Dan Punday, Mark Alleyne, Norman K. Denzin, James M. Salvo, Clifford G. Christians, Melissa Orlie, Eric T. Freyfogle, Arthur Melnick, Richard Schacht, Hugh Chandler, Avé M. Alvarado, and Steve Burkett. Thank you for your faith in me.

I am also thankful to my mother, Carmen S. Ortiz Aponte; my father, Daniel Vélez Correa; and my dear brother Daniel Vélez Ortiz for his unfailing love and support. To my colleagues in the Ethics Division of the National Communication Association, my everlasting requital. Lastly, thank you to Benny Herger, Adel Iskandar Farag, Eric Horn, Allison Payne, Manuel Aviles Santiago, Matthew Rukgaber, Ramón Soto Crespo, Rob West, Miles C. Coleman, and Adrianne Wallace, and my beloved Chad, who missed many long walks while I was researching and writing this project.

Excerpts from Fisher, Loren R. *The Eloquent Peasant.* (Cascade Books, 2015) used by permission of Wipf and Stock Publishers. www.wipfandstock.com.

1   The Evolution of Maat as the Collective Guiding Principle in Pre-Colonial African Civilizations

Figure 1.1 Wall. New Kingdom: XVIII Dynasty (before Amarna). The Ancient World, Egypt, New Kingdom, XVIII Dynasty, before Amarna.

In his seminal *A Sand County Almanac*, American ecologist Aldo Leopold describes ethics as "a kind of community instinct in the making" (203). This way of conceptualizing ethics lays bare the intentionality of ethical approaches as well as its communal dimensions. As an ecologist, Leopold is interested in the role ethics performs to achieve the end goal of human survival. At their core, most approaches to ethics tacitly or explicitly seek to do just that; promote the sanctity of life (Christians and Traber 6–7); and, by extension, prolong the survival of the human species. In Leopold's case, he connects both the survival and the flourishing of humankind to the survival of other animal species, landscapes, and *biotas*, a term that Leopold uses to describe the sum of components of local ecological communities (252).

Leopold's approach to ethics is rooted in American values of conservation, and social ecology but the idea that human flourishing and survival are tied to the health of the land can be traced many thousand years back to the first cultures that developed agriculture, domesticated animals, and other vital natural resource management techniques in the continent of Africa (Marshall and Hildebrand 100). The fact that tracing the evolution of Maat as an ethical view requires that we revisit the fundamental role ethics plays to extend, support, and valorize human survival on the planet may at first appear peculiar, but, in fact, it is faithful to the nature of ethical precepts and protonorms (Media Ethics 101). Contemporary advances in the fields of archeology, anthropology, geology, and climate research have rocked the foundations of ancient history on a global scale. The seismic changes in theories and new evidence continue to evolve as scientists document and verify previously unavailable data shaking the foundations of many long-standing beliefs and narratives about the lack of intellectual contributions made by classical African cultures. One such narrative is that African peoples lack a pre-colonial history. In this version of history, ancient Africans are nomadic, unsophisticated, superstitious, and largely undifferentiated people that all but needed to be civilized by Europeans. This colonial narrative has served a convenient exculpatory mission (The Ideology of Racial 1; Browder 34; Bernal 438; ben-Jochannan 1); it has placed Eurasians at the top of the social hierarchy and credited them with all moral, philosophical, and intellectual human achievements, while ancient African and other ancient civilizations such as the Maya, the Aztec, the Inca, and the Olmec are credited with technological and military accomplishments but no moral, rhetorical, or ethical ones. This way of understanding ancient history is no longer defensible. Instead, newly available evidence shows the story of rhetoric and, indeed, human communication ethics, doesn't start in Athens but in Africa.

Around 6000 BCE, the Nile Valley was first inhabited by people from what is now the Sahara Desert in middle Africa. The term Nile Valley is used to describe 4,100 miles of civilization or "the beginning or the birth of what is today called civilization" (ben-Jochannan 83). These Proto-Sahanran civilizations provided the pastoralist and agriculturalist (Heine 82) cosmological axioms in which later, even contemporary views, on ethics and human communication are based. The rhetorical and human communication view that arose from these first human civilizations is the one that is intimately tied to our inter-dependence on the natural environment. Such a foundation also yielded clear answers to vital ontological questions such as: What are humans for?; What is good?; What does it mean to be?; and epistemological questions such as: How do we know?; What are the limits of human knowledge?

In the 21st century, cosmology has become a less popular view of the nature of human existence in western societies, but it is still with us as it is still one plausible way to connect primordially human beings to the natural world. The fact that cosmology makes an appearance as the foundation for the first human ethic (or at least the first for which we have evidence) should not come as a surprise. In many ways, the ability of our species to survive on this planet and the necessary ability to adapt to harsh and never-ending climatological and other environmental conditions have been made possible through our keen observation of natural processes and patterns, and to our linking in meaningful ways of our individual and collective wellbeing to the physical and spiritual world.

The Earthshattering End of the Last Ice Age

Paleobotanical research estimates the Sahara Desert formed in the Pliocene (Micheel 193) and has alternated between lush and dry phases over the last few hundred thousand years, leaving regions uninhabitable and others newly fecund at different periods. Moreover, a 2013 study by Lasoarraña, Roberts, and Rohling has confirmed that tropical Africa has played a key role in the evolution and population dynamics of hominim by offering a better understanding of the biome habitability of the Sahara over a span of eight million years. The findings of their study confirm the previously estimated age of the Sahara by both archeological and fossil hominim records.

The reasons for the climatological fluctuations in the Sahara region lie in earth's orbital cycles. Perhaps, surprisingly, earth's gravitational interactions with the moon and other large planets have impacted human ethics and rhetorical practices through the human response to changes in earth's axial rotation during the Holocene epoch. Such interactions occur in a 41,000-year cycle because of a tilt in the earth's axis or "obliquity" that has been shown to have impacted African climate and African migration patterns by driving monsoon rains north and intensifying their strength (deMenocal and Tierney par. 3). The picture that emerges from this cursory overview of the orbital factors that forced subtropical temperatures during the last glacial period (deMenocal and Tierney par. 4) is the one that significantly alters evolving human values, priorities, culture, and human ingenuity in the quest for self-preservation. From an ethics and rhetorical perspective, the Green Sahara Periods (GSPs) become key sites of human interaction and it is during GSPs that successful hominid migrations, settlements, and social organization depend most heavily on the development of social norms and communication practices to coordinate adaptation to a wide range of paleoclimatological and paleoenvironmental events.

But what of the specific developments in human communication and rhetoric(s) during the holocene? Although the work of prehistory presents daunting challenges to those attempting to trace human development in ancient times, Anthropologist H.J. Hugot sums up the research into the cultural and technological achievements of Saharan civilizations following the last ice age as both the "cradle of cultivated plants" and "a centre for animal domestication" (Hugot 488). When we talk about ancient Saharan cultures, then we are honing in on the very people who through resourcefulness, imagination, and keen practices of observation made it possible for humans to inhabit a dry northern Sahara as well as the rain-rich southern region. As one scholar explains what the data now reveals about the importance of the continent of Africa in man's cultural development:

> REMARKABLE and exciting discoveries that have been made in Africa during the last five years suggest that it was here that tool-making first appeared in the geological record, and that it was then carried to other continents by hominid forms, the discovery of which has necessitated completely new thinking about the biological development of man. . . Therefore, it is now obvious that archaeology can provide some of the best source material for the reconstruction of cultural antecedents, population movements, and even of the origins of some social and religious practices on a factual basis, it is the new ways in which the archaeologist is using his data that render the results and potential as valuable (Clark 161).

As Norman Denzin reminds us in his canonical *Symbolic interactionism and cultural studies: The politics of interaction*: "The personal and the structural are mediated through the process of communication" (27). Symbolic interactionism is just one tool scholars have used to connect every day human interactions with objects and cultural practices to language use and the creation of meaning. There is a symbiotic relationship between what we do physically every day and the individuals that interact in such settings and this relationship is negotiated through language. As Denzin describes the cumulative result of said individual interactions: "The spirit of a society is transmitted and perceived through the cultural objects that it presents to its members" (143).

Hence, wild fluctuations in climate played a significant role in the development tools, in spreading around the use of these tools, and in forcing groups to migrate along the continent of Africa for many thousands of years prior to the emergence of dynastic Egypt. In this way, climate change played a role in bringing people of otherwise disparate groups into close contact, and in the exchange of ideas, values, and customs among quite diverse populations (Wengrow et al. 107). What emerges in speech from these interactions are cultures that put a high value on logic, in scientific speech, and a deep connection to the natural environment.

Take, for example, one of the main languages spoken in ancient Kemet: MDW NTR. Even if we set aside the question of whether NTR (divine, of God, God) is the root word for the word "nature" (Mayers-Johnson 154), we still have to recognize that MDW NTR (or hieroglyphs as it is often called after the Greek words for sacred carvings (Ritner 74) is characterized by a consistent and signature representation of the elements of nature including animals, plants, and landscapes. This was intentional. To these speakers, it was far more efficient to communicate using images that convey multiple thoughts than using grouped letters. By relying on familiar images of animals and other aspects of nature, rhetors placed emphasis not on the image itself as one would in worship, but instead on how the particular animal, plant, or landscape behaves in particular situations, all of which are aspects of the divine (Nihusi 364). Thus, the focus of the MDW NTR language as a manifestation of an overarching Maat ethic is on thoughts not words as a Greek form of persuasion would. As Egyptologist Jan Assmann notes:

> For what the heart thinks up are not the names of things but their "concepts" and their "forms." Hieroglyphic script is a pictorial rendering of the forms. It relates to the concepts by way of those forms. The tongue vocalizes the concepts, which were "thought up" by the heart and given outward and visible form by hieroglyphic script (27).

A Genealogy of Morals

To date, the specific origins of Maatian spiritual or ethical views are unknown. In the previous section, I made a case for Maat growing and developing out of a pastoralist worldview that evolved through many earlier African cultures and became formalized and personified in Kemet. Of course, formally advancing such a claim would require

a much deeper look at the prehistoric civilizations of Africa emphasizing the spiritual beliefs of said civilizations. For the purposes of the
present inquiry, my task is considerably smaller. In highlighting the
agrarian achievements and values of earlier African civilizations and
showing the continuation of that worldview in Kemetic Maatian practice, I hope to show that such a connection is at least likely.

An ethical theory is "a systematic exposition of a particular view
about the nature of good or right" (MacKinnon 19). These theories
provide explanations or principles for judging individual acts to be
right or wrong and attempts to justify these norms. According to
Hugh Dalziel Duncan, in his essay "Communicative bonds and moral
bonds," the problem of morality in communication raises two questions: "what ought we to communicate" and "how do we communicate
what we ought" (73). Notice that, while a more general conception of
communication ethics may assess the effectiveness of communicating messages through various means (media), moral communication focuses on the study of communication about values. German
Sociologist Thomas Luckmann (1997), in his *Moral Communication
and Modern Societies*, not only agrees with this basic description of
the field but adds that the subject of moral communication unfolds
in socially constructed sign systems, particularly, but not exclusively
in language (2). This insight can be traced back to the classical ethical communication tradition, including ancient Pan-African views,
which posited that, for human beings, the "lingual interpretation of
ourselves and our experience constitutes who we are. Thus, human
communication and action are dialogic" (Keeble 64).

To be sure, content/form distinctions take center stage in the questions suggested by Duncan. However, far from remaining at the level
of description, the questions asked by scholars of moral communication aim to scrutinize what ought to be the case for moral action and
how this action can be aided by communication practices. Par for the
course, then, is to make normative judgments about desirable and undesirable, right and wrong, moral and immoral, form/content articu-
lations in communicative processes. Luckmann defines morality as a
"reasonably coherent set of notions of what is right and what is wrong,
a set of notions about the good life that guide human action beyond
the immediate gratification of desires and the momentary demands of
a situation" (1). Luckmann also explains that values "are constructed
in long historical chains of communicative interactions, and they are
selected, maintained and transmitted in complex social processes" (1).
Values, then, are dependent on communication processes for their formation, interpretation, and transmission. Duncan makes this point

more sweepingly by stating that "the social and the symbolic arise in communication" (76).

As an approach to ethics, Maat is a unique and powerful example of how the relationship between ethics and communication is inextricable. For as long as humans have thought about ethics as end goal, they have theorized communication as its corresponding means. In the Kemetic view of communication, the word, speech, and language itself are conceived as divine. Kemites thought of justice, reciprocity, and balance as end goals for human existence (in this mortal realm and others) and they thought of communication as the ideal means or vehicle by which to achieve these goals. For instance, the word *heka (hk3)* in Medu Netcher meant "word" but it was also associated with magic (Wise 21). *Heka* represents the creative force that brings the world from a state of chaos to order; it is the very act of creation that is available to human beings in a smaller scale through every day speech. This is a time in human history when our words were truly our bond and as one scholar puts it: "deeds did not necessarily speak louder than words. They were often one and the same... Thought, deed, image, and power are theoretically united in the concept of *heka*. The world is created with, through, by, and for speech" (Wise 21).

In this context, there is no (good or excellent) communication without ethics and no ethics without communication. Moreover, as we have seen, Maat is not a belief system that formed overnight or can be thought of as the exclusive property of Kemet. Instead, a number of religious views and cultural aspects, including the language of Medu Netcher, likely originated in Kush land (on the southern border of Kemet, modern-day Sudan) and perhaps other parts of the continent of Africa. Evidence of this is the 1962 discovery of what is potentially the oldest Nubian monarchy in the old city of Ta-Seti of ancient Nubia. To presume that ancient Kemet sprung out of thin air is a mistake that has plagued many early Egyptologists (Ben Levy 178). Modern research has shown that Kemet was in direct contact and communication with cultures worldwide and most certainly shared ancestry, language, and customs with other African civilizations, both contemporaneous and preceding Kemet (Connor 2; Makki 161).

Ta-Seti is the oldest recorded kingdom in the world. Medu Neter is the oldest writing system in human recorded history. That is why you have to have maps and time charts dealing with African people. African people already had nations 4,500 years ago (Hilliard 199).

In short, Maat, or at least, the foundations of Maat likely come from through a much longer trajectory of pastoralist and agrarian cultures that preceded the emergence of the dynastic Egypt. As one scholar puts it:

> Unlike in Classical Greece, Egyptian cosmological, metaphysical and ethical concepts did not crystallize over the course of a few centuries. Instead they were the outcome of *millennia* of intellectual labour, during which hundreds of priests developed and grappled with challenging, often contradictory, ways of making sense of the universe (28).

While the exact origins of Maat await discovery, there are many clues to its scope, reach, prescriptions, and protonorms. For instance, the core or fundamental commitments of Maat as an ethical perspective can still be found in the lexicon and value systems of contemporary African societies: "Ma'at, as a term and principle, survives among many African populations whose languages belong to the Cyena-Ntu (C-N) language family (Imhotep 1)". We also know that the values promoted by Maat extend the geographical boundaries of modern-day Egypt, as they probably did in ancient times. We also know that Maat was an overarching ideal that was used as a guide spiritually, politically, and philosophically in ancient Kemet. Given its hold on Kemetic society and its reach into neighboring cultures, perhaps it should not be surprising that even today, "Maat is the central principle of governance throughout all of Africa and truly is, indeed, a Pan-African cultural framework" (Imhotep 18).

Ecological Dimensions

Another vital aspect of Maat as an ethical approach is its connection and emphasis in our human inter-dependence with nature. Kemetic society maintained and promoted an attitude of kinship, reciprocity, respect, and deep curiosity for all cosmic substances including those found in nature. Today, we might associate the unseen, hidden (occult), what lies beyond the surface as something of a scary proposition with some sinister implications, but the Kemetic mind thirsted for knowledge and located truth precisely in that which requires disciplined study, keen observational skills, and logic in order to adduce and deduce truths. In the words of rhetorician Edward Karshner, "The power of *heka* was found in its close association with language. In fact, the whole of Egyptian mysticism and magic is encoded in the metaphysics of its linguistics (56)".

A cursory look at a few of the symbols commonly used in the Medu Netcher (Divine Speech) language can provide insight into the sort of objects, attributes, and earthly creatures that fascinated Kemites and captured their imagination. Immediately, one can't help but notice that many of these symbols reference animals, landscapes, and tools for which Kemites felt appreciation or found especially useful in their popular crafting endeavors. Much like a craft, Kemites thought of speech as a powerful creative force (*Heka*) that invested moral agents with the ability to create reality, indeed like magic, through both action and deed. It is important to emphasize that for ancient Kemites, it was not only possible to speak Maat but do Maat. Moreover, a highly developed interest in the natural world and its inhabitants (human and non-human) appears to have shaped their language structure. In his study of the Medu Netcher language Asar Imhotep has found some basis for the idea that the Greek word "logos" first introduced by pre-Socratic philosopher Heraclitus may be connected to the way in which Kemites understood wisdom and logic.

The Medu Netcher word for logos is Maat. Maat represents "the word" just as Medu Netcher translates to divine or sacred speech, and *heka* means "word" and it is also related to power, and the prinoridal creative force. According to Imhotep, the Medu Netcher language is built on the foundational concept of a tree. From a tree, we get the concepts and words for the human body (trunk-like) which then evolves into conceptualizing bodily appendices such as arms and legs as branch-like. This is a naturalistic view of the human experience that is free from any semblance of a human/nature dichotomy. In the table below, Imhotep looks at various basic lexical units (lexemes) in contemporary African languages and finds that as many African languages go through phases, including Medu Netcher, the concept of lége appears connected to the concepts of foot, leg, or member which, at later stages, becomes associated with "rightness" "walking on the right path" and later with logic or logos itself (Table 1.1).

Not only is Medu Netcher (Divine Speech) comprising naturalistic ideograms "but the word Ma'at is also the word for community" (Imhotep 32). The picture that emerges is the one we might recognize today as ecological in terms of a shared preoccupation of how the individual fits into the natural environment. Kemites thought about community in much broader terms than simply the human individuals that comprise a city or a nation. They were thinking about the inter-dependence that existed between themselves and the land (with all its features) that they were grounded in.

Table 1.1 The evolution of *m3ʿ*

Lexeme	Mdw-nTr	Meaning	Source	tree > branch >
m3ʿ	*(hieroglyphs)*	"wood, board, plank"	Vygus 1862	stick, object
m3ʿ	*(hieroglyphs)*	"place, court of a house or temple"	Budge 272b	member > foot > place >
m3ʿ	*(hieroglyphs)*	"shore, bank of a river, flat near mouth of a river"	Budge 272b	member > foot > walk > footpath > road
m3ʿ.w	*(hieroglyphs)*	"a promenade by the river(?)"	Budge 272b	member > foot > movement > walk > footpath > road
m3ʿ.w	*(hieroglyphs)*	"to go, to journey, to go straight to a place"	Budge 272b	member > foot > movement > walk, run
m3ʿ	*(hieroglyphs)*	"salt water"	Budge 272b	member > foot > movement > stream, river
m3ʿm3ʿ	*(hieroglyphs)*	"to go, to travel" (reduplication)	Budge 272b	member > foot > movement > walk, run
m3ʿ.w	*(hieroglyphs)*	"advance guard, pioneers, soldiers"	Budge 272b	member > foot, leg > movement >
m3ʿ.w / m3ʿ.wt	*(hieroglyphs)*	"to sail, wind, breeze"	Budge 272b	member > foot > movement >
m3ʿ.w	*(hieroglyphs)*	"cordage of a boat"	Budge 273a	hardness > force > magic > animal > snake > rope
m3ʿ.w	*(hieroglyphs)*	"hook, clasp"	Budge 273a	hardness > force > spine (peak) > hook[a]
m3ʿ	*(hieroglyphs)*	"eyebrow"	Budge 273a	member
m3ʿ.tj	*(hieroglyphs)*	"the temples of the head"	Budge 273a	member
m3ʿ	*(hieroglyphs)*	"to kill, to slay"	Budge 273a	hardness > force > strike > kill
m3ʿ	*(hieroglyphs)*	"boat"	Budge 273a	member > foot > movement > stream, river > water > canoe

a Given that this term has a classifier (determinative) of a tree branch, this semantic evolution may also fall under: *tree > branch > stick, object > thing*.

Today, ethical views that are based on our embeddedness into nature are considered to be ecological. Known as the father of ecology, Aldo Leopold claims that "a thing is right when it tends to preserve the integrity, stability, and beauty of the biotic community". This sentiment is very close to that original Maatian quest for harmony, for balance, and for desire to walk the right path that leads to harmony. Kemites knew that they were blessed by the annual flooding of the Nile as it would imbue the soil with the nutrients needed to sustain life in that landscape all year round. It should be no surprise, then, that their ideas about rightness and wrongness have some connection to the life-enabling resources that allowed life to continue and prosper along the river Nile.

Ecologically, "a thing is right when it tends to preserve the integrity, stability, and beauty of the biotic community" and with this simple moral maxim, Leopold describes both a contemporary scientifically informed compass with which to properly appraise the rightness or wrongness of the human relationship to the land and an ancient Kemetic view of ethics.. For Leopold, the idea of land health is imbued with the concern for preserving the self-renewing capacities of the land so that it may sustain the greatest level of biodiversity and facilitate the survival of its citizens, including, of course, human beings. The end goal then of a proper land ethic and a Maatian ethic is a "shared duty" that reflects "the existence of an ecological conscience, and this in turn reflects a conviction of individual responsibility for the health of the land" (Leopold 22).

The fact that Kemites made such extensive use of animal imagery in painting, sculpture, carvings, and Medu Netcher itself has led many Egyptologists in the past to erroneously attribute animal worship to the Kemites. This interpretation has also advanced the myth that Kemet was a nation steeped in superstition and ignorance. Indeed, the Medu Netcher is filled with images of birds, mammals, and other species but one must not conflate admiration for worship. It is one thing to see oneself as being in the service and whims of a being (of whatever species) and another to express admiration and inspiration in its attributes.

In sum, the aspirations of a land ethic and a Maatian ethical worldview can be summarized in very similar ways: Both shared a linguistic recognition of inter-dependence, and imperative of self-knowledge (including one's limitations), knowledge of the land (as required by fluency in Medu Netcher), as well promoting virtues of compassion, enthusiasm, curiosity, and affection complimented by action guided by these ideals. Remember, in Kemetic society, one was prompted to speak Maat and do Maat. All of which, as features of an ethic, are quite complimentary though not identical.

Like a contemporary Leopoldian conservation ethic, Maat deals in connecting and emphasizing values such as health and communities as vital aspects of our relationship to land. This is because both ecological ethics and Maatian ethics discourage the idea that, "we are free to discard or change any part of the land we do not find 'useful' or intelligible for that matter (such as flood plains, marshes, and wild floras, and faunas)" (The land we share 22). For Leopold, the idea of land health was meant to reflect the dynamic and broad set of contexts in which conservation can provide ethical norms in trying to adjudicate between the one-and-the-many:

As Kemites would agree, Leopold described healthy land as "stable". Not to suggest that natural systems are static but in the more specific sense that land retains its ability to cycle nutrients effectively, and thus maintain its soil fertility. In order to do so, "the land needs to have integrity, by which Leopold meant the biotic parts necessary for this nutrient cycling to take place. Leopold uses "stability" and "integrity" in tandem as a shorthand expression for "land health" (Why is conservation 23).

Of course, land health depends on the knowledge that human beings can gather (and indeed articulate to one another) in order to understand the various relationships of inter-dependence that are at play in any given eco-system. This is something Kemites built into Medu Netcher, and something that was built into the 42 neteru (divine principles of wisdom, forces of nature) who in Kemetic cosmology represented an aspect/force of nature (Mercer 12). However, it is critical to make the distinction that while Leopold did not scoff at the insights provided by a scientific knowledge of the land, the type of knowledge that he is endorsing in the previous quote is more akin to a practical, hands-on knowledge that mirrors the Kemetic approach to ecology. Leopold referred to this type of knowledge simply as skill and skill for Leopold came from:

A careful attentiveness to the land and from a readiness to respect nature's equal management role. Skill arose within a person possessed a lively and vital curiosity about the workings of the biological engine, a person inspired by "enthusiasm and affection." These were the human qualities requisite to better land use (Why is conservation 90).

Conclusion: Maat as a Land Ethic

In contemplating Maat as an overall conservation goal, akin to Leopold's land ethic, we must appreciate the role of virtues such as enthusiasm, curiosity, affection, and knowledge as cultivated dispositions vital to this approach to ethics. Both the Land Ethic and in Maat are premised on the principle that we have inescapable claims on not just one another but on the land, which cannot be renounced except at the cost of our humanity. In contrast to a Leopoldian land ethic, a Kemety cosmology yields a view of nature and God as one and the same. Kemet was a society in which to do that which affronts the divine rose to the level of *isfet* (evil). Such a view of rightness and wrongness is part of a broader legacy of knowledge and values developed over many thousands of years by civilizations that preceded Kemet and engaged in a pastoralist existence that far from immersed in ignorance, flourished, and made great advances in mathematics, astronomy, philosophy, and yes, ethics. The ultimate goals of Maat are balanced, harmony, reciprocity, justice, and truth for posterity. These are indeed demanding and ambitious goals and thus the acceptable means for living the good life for Kemites was equally as demanding in terms of holding each citizen to a high standard of virtue in their quest to become "true of voice" in the future life (often referred to as afterlife). In this way, Maat offers a unique mixture of cosmology, ecology, and spirituality that is overdue to be acknowledged as the prototype for all other subsequent approaches to human (communication) ethics.

Bibliography

Asante, Molefi K. *African Intellectual Heritage: A Book of Sources.* Temple University Press, 1996.

———. "The Ideology of Racial Hierarchy and the Construction of the European Slave Trade." *Black Renaissance/Renaissance Noire*, vol. 3, no. 3, 2001, pp. 133.

Assmann, Jan. "Creation Thought Hieroglyphs: The Cosmic Grammatology of Ancient Egypt." *The Poetics of Grammar and the Metaphysics of Sound and Sign*, edited by Sergio La Porta and David Dean Shulman. Brill, 2007, pp. 17–34.

Barton, George A. "The Origins of Civilization in Africa and Mesopotamia, Their Relative Antiquity and Interplay." *Proceedings of the American Philosophical Society*, vol. 68, no. 4, 1929, pp. 303–312.

Ben Levy, Josef Y. "The Intellectual Warfare of Dr. Jacob H. Carruthers and the Battle for Ancient Nubia as a Foundational Paradigm in Africana Studies: Thoughts and Reflections." *The Journal of Pan African Studies*, vol. 5, no. 4, 2012, pp. 178–195.

Christians, Clifford G. "The Ethics of Being in a Communications Context." *Communication Ethics and Human Values*, edited by Clifford G. Christians and Michael Traber. Sage Publications, 1997, pp. 3–23.

———. *Media Ethics and Global Justice in the Digital Age.* Cambridge University Press, 2019.

Clark, J. Desmond. "The Prehistoric Origins of African Culture." *The Journal of African History*, vol. 5, no. 2, 1964, pp. 161–183.

Connor, David O. "Ancient Egypt and Black Africa- Early Contacts." *Expedition*, vol. 14, no. 1, 1971, p. 2.

deMenocal, Peter B. and Jessica E. Tierney "Green Sahara: African Humid Periods Paced by Earth's Orbital Changes." *Nature Education Knowledge*, vol. 3, no. 10, 2012, p. 12.

Denzin, Norman K. *Symbolic Interactionism and Cultural Studies: The Politics of Interpretation*. Blackwell, 1992.

Duncan, Hugh D. "Communicative Bonds as Moral Bonds." *Communication: Ethical and Moral Issues*, edited by Lee Thayer. Gordon and Breach Science Publishers, 1973, pp. 73–96.

Flegel, Peter. "Does Western Philosophy Have Egyptian Roots?" *Philosophy Now*, vol. 128, 2018, pp. 28–31.

Freyfogle, Eric T. *The Land We Share: Private Property and the Common Good*. Island Press, 2003.

———. *Why Conservation Is Failing and How It Can Regain Its Ground?* Yale University Press, 2006.

Hayward, Richard. "Afroasiatic." *African Languages: An Introduction*, edited by Bernd Heine. Cambridge University Press, 2000, pp. 74–98.

Hazen, Robert M. *The Story of Earth: The First 4.5 Billion Years: From Stardust to Living Planet*. Penguin Books, 2013.

Hilliard A. G. "The Teachers of the World." *How Long This Road. Black Religion/Womanist Thought/Social Justice*. Edited by Alton B. Pollard and Love H. Whelchel. Palgrave Macmillan, 2003, pp: 195–202.

Hugot, J. H. "The Origins of Agriculture: Sahara." *Cultural Anthropology*, vol. 9, no. 5, part 2, 1968, pp. 483–488.

Imhotep, Asar. *Aaluja Vol. II: Cyena-Ntu Religion and Philosophy*. Madu-Ndela Press. 2020 (Forthcoming).

Karshner, Edward. "Thought, Utterance, Power: Toward a Rhetoric of Magic." *Philosophy & Rhetoric*, vol. 44, no. 1, 2011, pp. 52–71.

Keeble, Richard. *Communication Ethics Today*. Troubador Publishing, 2006.

Khan, Abdul J. *Urdu/Hindi: An Artificial Divide: African Heritage, Mesopotamian Roots, Indian Culture & British Colonialism*. Algora Publishing, 2006.

Leopold, Aldo. *A Sand County Almanac and Sketches Here and There*. Oxford University Press, 1949.

Luckmann, Thomas. "Moral Communication in Modern Societies." *Human Studies*, vol. 25, no. 1, 2002, pp. 19–32.

Mackinnon, Barbara. *Ethics: Theory and Contemporary Issues*. Wadsworth Publishing, 2003.

Makki, Fouad. "The Spatial Ecology of Power: Long-Distance Trade and State Formation in Northeast Africa: The Spatial Ecology of Power." *Journal of Historical Sociology*, vol. 24, no. 2, 2011, pp. 155–185.

Marshall, F. and Elisabeth Hildebrand. "Cattle before Crops: The Beginnings of Food Production in Africa." *Journal of World Prehistory*, vol. 16, no. 99, 2002, pp. 99–143.

Mayers-Johnson, Sheila. "The Word as a Conduit for African Consciousness: Mdw Ntr through Ebonics." *Journal of Culture and Its Transmission in the African World*, vol. 1, no. 2, 2004, pp. 145–176.

Mercer, Samuel A. B. *Growth of Religious and Moral Ideas in Egypt.* Morehouse Publishing, 1919.

Micheels, A., J. Eronen, and V. Mosbrugger. "The Late Miocene Climate Response to a Modern Sahara Desert." *Global and Planetary Change*, vol. 67, nos. 3–4, 2009, pp. 193–204.

Nihusi, Kimani S. K. "Humanity and the Environment: Environmentalism before the Environmentalists" *The African Union Ten Years After: Solving African Problems with Pan-Africanism and the African Renaissance*, edited by Mammo Muchie, Phindil Lukhele-Olorunju, and Oghenerobor Akpor. Africa Institute of South Africa, 2013, pp. 364–382.

Ritner, Robert K. "Egyptian Writing." *The World's Writing Systems*, edited by Peter T. Daniels and William Bright. Oxford University Press, 1996, pp. 73–83.

Semaw, Sileshi et al. "Early Pliocene Hominids from Gona, Ethiopia." *Nature*, vol. 433, no. 7023, January 20, 2005, pp. 301–305.

Weaver, Timothy D. "Did a Discrete Event 200,000–100,000 Years Ago Produced Modern Humans?" *Journal of Human Evolution*, vol. 63, no.1, 2012, pp. 121–126.

Wengrow, David et al. "Cultural Convergence in the Neolithic of the Nile Valley; A Prehistoric Perspective on Egypt's Place in Africa." *Antiquity*, vol. 88, no. 339, 2014, pp. 95–111.

Wengrow, David. *What Makes Civilization? Ancient near East and the Future of the West.* Oxford University Press, 2010.

White, Tim D. et al. "*Ardipithecus Ramidus* and the Paleobiology of Early Hominids." *Science*, vol. 326, no. 5949, 2009, pp. 64–86.

Wise, Christopher. "nyama and Heka: African Concepts of the Word." *Comparative Literature Studies*, vol. 43, no. 1/2, 2006, pp. 19–38.

2 Ancient African Spirituality
Heaven on Earth

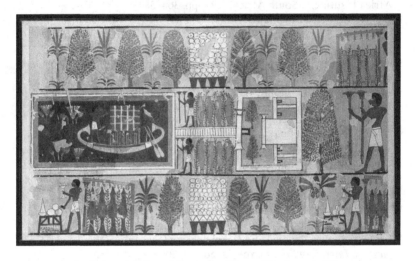

Figure 2.1 Egyptian; Thebes 18 Dynasty; Time of Tuthmosis III.

It would be a mistake to confuse a robust spirituality for a formal religion, especially in the context of ancient Kemet. In the previous chapter, we discussed some of the ways in which Kemites saw themselves as embedded in nature and expressed their kinship to it in a variety of ways, including through the language of Medu Netcher. The aforementioned image, found in the tomb of an overseer of granaries (Wilkinson 12), illustrates the motif of beautiful Theban gardens and even a pool so generous in size that a boat is able to sail in its waters. Like many Kemetic artistic images, human beings are shown in the midst of a lush and expansive landscape. Depictions such as this offer insight into the reaches of the Kemetic imagination, to what was valuable to them, and where their minds wondered when thinking about transporting themselves. Perhaps it comes as a surprise that in many instances, the Kemetic imagination, through art, speech, and science, traveled with gusto and inspiration to none other than the land of Kemet.

We can hardly blame the 19th-century European archeologists who traveled great distances to be enchanted by the exotic lands of the "orient", for equivocating the lavishly decorated tombs they found for a culture of death (Lant 93; Shaw 62; Park 530), preoccupied with escaping the mortal realm to a mysterious underworld (*Duat*) and afterlife, as opposed to the life-loving, life-affirming, and nature-loving people Kemites truly were. Furthermore, the presence of deities and iconography may appear to the untrained eye as an obvious sign of an active formal religious practice but to do this would be to impose a European interpretation of these facts while ignoring the particularities of long-standing cultural systems of belief found in many other parts of Africa (Marumo and Mompati 11700). Regrettably, this is exactly what has transpired in the field of Egyptology since its inception. As one scholar puts it:

> it created not only a new form of knowledge, Egyptology, but also new ways of understanding the imperial power relations underwriting this European science, perhaps the first academic discipline whose fortune wholly depended on colonial domination. In fact, the new science provided a confidence in interpretation that was not just limited to ancient artifacts: arguments about the superiority of European sciences of interpretation were crucial in developing a wider, colonial sensibility that it was Europeans, not Egyptians, who knew Egypt best (Colla 76).

The power dynamics and political implications woven into the fabric of the field of Egyptology cannot be underestimated. Those first accounts, interpretations, and translations of the artifacts found in Kemet still resonate and inform the corpus of knowledge of Egyptology. Take, for instance, the idea that in order for a society to prosper and organize itself in a meaningful manner, its people must adopt some form of religious practice. Let's also keep in mind the European Egyptologists who inaugurated the field that came from societies in which large-scale religious conflict had marked some of the most important moments in their history. The idea that an advanced civilization could have made the sort of achievements Kemet produced without organized religion would've been almost unthinkable.

Yet, that is exactly what not just in Kemet, but other advanced pre-colonial civilizations in Africa did.

> It is important to note that the African God, at the level of reve-lation, was the same in essence as the Christian or Muslim God introduced into the continent by missionaries. . . . They differed in specific doctrines and beliefs, however, e, g, the need for a specific building in which to worship God, belief in a heaven and hell, in the afterlife and denominalism, among others. In short, the es-sence of the African belief was spirituality as opposed to organ-ised religion (Malunga 12).

The Communicative Dimensions of African Spirituality

Although often grouped, there are qualitative differences between the concepts of religion and spirituality that merit examination. First, there is a difference between admiration and worship. To admire the attributes of a thing is qualitatively different from worshipping said thing or attributing to it a supernatural quality. To unpack this con-trast, let's start by ruminating on the concept of natural as half of the notion of the supernatural. What is natural has indeed been highly dependent on cultural values, beliefs, and available science through-out history. From natural law, to natural history, to natural histories of disease, the concept of what is "natural" continues to evolve as do cultures worldwide. Furthermore, we can trace the displacement of human beings from the realm of nature (or the natural) to the enlightenment period with the man/animal separation made by Renee Descartes in his *Meditations*. Aristotle conceived of human beings as "speaking animals" and this lack of dissociation between the man and the rest of nature is the closest chronologically and philosophically to the Egyptian view of nature than Descartes'.

Having explored some ideas about the "natural" related to our con-cept of supernatural, it is worthwhile to inquire about what is "su-per" or extra-natural. This is of special importance because one of the claims I am advancing in the current analysis is that the Kemetic cos-mology that produces Maat does not have such a concept. Evidence of this can be found in the very presence of the neteru in Kemetic cosmology as forces of nature. In this version of spirituality, what is considered to be the highest and mightier than all other things is not apart, extraneous, or separated from the cosmos, from nature. On the contrary, the neteru are the fulfillment, the fruition, the fully realized versions of what all human beings are in possibility. This is reminiscent of Aristotle's own version of metaphysics in which potentialities, in this case human beings, function as the potential of what a neteru has made actual. In Aristotle's own words:

> In the case of matter-form compounds and numbers the actuality is accompanied by potentiality—perceptual sublunary matter in the first case, intelligible matter in the second. In the case of divine substances and other such unmoved movers, it is not. They are "pure" activities or actualities, wholly actual at each moment (xxxiv).

In other words, there is nothing super or foreign about such forces of nature in Kemetic thought. The power of the neteru might be unseen or hidden from the untrained eye, but neters are composed of the very same substance (stuff) of which the whole of the cosmos is comprised not extrinsic to it. In other words, the entire cosmos is made from the same recipe and same batch. Neteru, then, rather than gods represented the unseen forces (infraworld) and divine principles of wisdom in the world (West 240) as well as the power that comes from their exercise (Lambert 46). Kemet was a place where there was a sacredness to everything, including and especially human beings precisely because they shared in the components of the divine. Just like neteru (neters), both male and female forms are made of the substance with which the cosmos is made. Thus, humans were viewed as part, highly dependent, and inextricably tied to each other, to the land, the air, the water, and every other cosmic substance.

Similarly, it is crucial to understand ancient African spirituality in its proper context and on its own terms. To gloss over the distinction between religion and spirituality would lead to a misunderstanding of the uniqueness of Maat as an ethical and communicative ideal. One scholar explains the difference between religion and spirituality in this way:

> In what the European calls religious; terms (the African needs no separate or distinctive category for this experience, since the universe is spiritual in nature), the cosmic order is sacred and our concern becomes the discovery and maintenance of that sacred order through proper relationship to the whole. A cosmology is the systematic explanation of the interrelationship, origin, and evolution of natural forces and of the nature of those forces. This includes man's relationship to nature and the significance of that relationship. Mythology involves the personification of this

explanation. Traditionally, Africans understand their societies as sacred in origin. This understanding helps to sacralize the community. Cosmological and mythological systems are therefore elaborate, complex, and profound, since they are crucial to the explanation of the sacredness of these origins, from which the present derives meaning and significance (Richards 219).

Another team of scholars put even more pressure on the ritualistic and attitudinal differences between religion and spirituality in an African context:

> There is a difference between spirituality and religion. Religion is the service and adoration of god or a god expressed in forms of worship. Religion refers to an external formalized system of beliefs whereas spirituality is concerned with a personal interpretation of life and the inner recourse of the people (Laukhuf and Werner 61).

It is the latter part of this quote that unlocks the power and true character of Maatian ethics. Although the ideal of Maat was personified and illustrated through the image of the winged woman figure, the power of Maat lies not on the mercy or the pity of the so-called "goddess" herself but in the way in which individual speakers and doers choose to behave, choose to act, and speak to others. Indeed, it is "the inner recourse of the people" that makes Maat an ethical ideal and not a religious sect. For all ethics must come from the inside. The difference between law and ethics is that laws are imposed externally and individuals are expected to comply with these laws that they did not create nor made into law. Ethics, however, comes from within; these are the laws that individuals give themselves and act upon only through their individual will. No one can make another person ethical; that would be a contradiction in terms. In a communication context, one can speak Maat if one chooses to do so and assuming such individual understands the benefits or goodness of choosing to speak Maat and they have willed to engage in that practice. In simple terms, participation in communicative and material ethical practice must be freely chosen and self-willed.

The broader point I'm making here is that Kemet's vigorous spirituality is a natural fit for ethicality. Thus, to imply that a lack of formal religious practice forecloses the possibility of a society being able to develop a proper approach to ethics is unreasonable and not supported by the evidence thus far discovered. Indeed, ancient Kemet did not have a formal religious practice. As Egyptologist Stephen Quirke states, there was no bible, no creed, nor a fixed system of beliefs (Quirke 2). What we do have available to examine their values, beliefs, and ethics is the wisdom literature of Kemet, and among them, the *Instructions of Ptahhotep*, written between 2375 and 2350 BC, as evidence of the presence and hold of the moral ideal of Maat in the Kemetic way of life (Karenga 266).

It isn't until the 1st century AD that Christianity makes its appearance (Case 4) the type of institutionalized religious order that we can link to our current practice. What Kemites had was a complex and fully realized spirituality that took root in all areas of society, both sacred and secular. While the Egyptomania industry places all of its emphasis in the numerous Egyptian "Gods" of yesteryear, a more accurate description of these figures lies in their *Medu Neter (mdw ntr)* language designation as *ntr*. An ntr is a specific manifestation of the divine (West 240). In the word NTR, T stands for feminine, while R stands for the masculine principle.

The divine, in turn, is the sacred or that which is to be most revered, promoted, and exalted intellectually and spiritually. For example, the *ntr* Thoth embodies the qualities of knowledge, judgment, wisdom, and reasoning that ought to be prioritized, encouraged, and popularized among human beings and appreciated in all creatures that possess it. Seshat, the *ntr* of writing, archives, and libraries, is no less an *ntr* than Thoth; in fact, she is his partner, and her sphere of appreciation and promulgation also very much takes place right here on earth, and is carried out by human hands, human minds, and hearts.

Thus, to view *neteru* in the contemporary fashion in which we conceive of gods misses the unique feature of Kemetic spirituality that kept one eye on the unseen, spiritual forces around us and the other on the myriad of ways human beings are the vehicles, the instruments, the very embodiment of the divine. Similarly, it is an error to interpret the locus of the divine outside of our cosmos. If anything, Kemetic people gave us indication after indication that the dwelling of the divine is right here in our milky way, in our planet, in our land, and in ourselves. Evidence of this can be found in the very origin myths circulating in ancient Kemet in which creation is said to have been formed from *Nun* (Kimkeit 280), or primeval water. According to two of them, the Memphite Theology and Heliopolis version, it was Ptah, the divine creative force (Ward 154), who not only created the *ntr* Atum through speech, but in doing so created the primordial order and harmony the *ntr* Maat is invested with promoting (Teeter 16).

In this cosmology, *Ptah* is neither male nor female but *Nut* (sky), and the celestial divine force is presented as female. Her role in the story of creation is of the greatest importance as it is she who contains the chaotic waters of *Nun* and imposes order upon it. From an empirical perspective, it is easier to appreciate the wisdom behind this story if one happens to live near oceans or large body of water and is aware of the effect celestial substances, especially the sun's energy, can have on water cycles. *Nut* is also said to represent the milky way (as seen in the image below) and her presence is often depicted as enveloping human experience and even human psyche all the way from seasonal depression disorders to lunacy (Cunningham 1950).

In short, in reviewing some of the cosmologies in circulation in Kemet, one can't help but notice that the Kemetic system of belief was very much circumscribed to our own cosmos, our galaxy, and our forces of nature. Even in myths where one would find allusions to the extra-natural, we see the Kemetic mind firmly focused on their time and place, their surroundings, and their world. Thus, to claim that Kemites were devoted or obsessed with anything other than this world is incorrect. Rather than setting their gaze on a different place, Kemites found a way to see their own milieu in inquisitive and integrated ways that made the soil underneath their feel take on an infinite number of connections to every aspect of their lived experience.

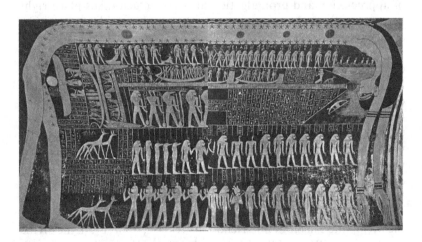

Figure 2.2 Ceiling. New Kingdom; XX Dynasty; The Ancient World, Egypt, New Kingdom, XX Dynasty.

From Cosmology to Communication Ethics

What is the connection between what happens in the cosmos and what happens here on earth? Djeuti (Thoth), the ntr (power) of wisdom and knowledge, who later the Greek renamed Hermes Trismegistus, imparted the doctrine of correspondence: "as above so below" (Robertson 406). This is a central tenet of Maatian ethics. In order to preserve, extend, and flourish as human beings, we must replicate the conditions of substances and forces that surround us. In a sense, no doctrine can get closer to Charles Darwin's notion of the survival of the fittest or the survival of those species who best conform or suit their material conditions. In the origin stories of Kemet (KMT), it was a feminine force that ordered the world from chaos to harmony and in the spiritual world the task of illustrating the virtues that epitomized good character also fell to a woman figure: Her name was Maat (also known as Ma'at or Mayet).

Depicted in the aforementioned image with her wings outstretched and in a kneeling humbled position, the ntr Maat was personified as the daughter of the ntr *Ra* (Robins 106). Variations of the aforementioned image abound across Kemetic artforms. While many Egyptologists and scholars have referred to this ntr as a goddess, she represents the power of speech, the power of truth, the power of harmony, the power of logic, and as she is personified, her power is being shown as available to those who practice her seven principles: "righteousness, truth, honesty, propriety, harmony, order and reciprocity" (Lynn 606).

Figure 2.3 The Godess Ma'at--2Am.128.

When the Kemetic people thought of neteru, they focused on principles of divine wisdom that are vital to preserving, extending, and enriching human life. The fact that these neteru are often depicted with partial animal features or as male or female evoked not prostration (as a god might evoke) but a visual reminder of the similarities between how nature behaves in particular situations and how that behavior represents specific aspects of the divine, of the cosmic (Ntwadumela 103). Perhaps one helpful way to grasp this concept is to think of *neteru* as principles of the sources of creation, aspects of creation that went on to be considered cosmic virtues. A cluster of such virtues can be seen in the representation of Maat. These are all virtues that at the time would have been considered conducive to maintaining societal order, peace, and by extension would prolong and promote human survival and prosperity. This observance later became codified into 42 declarations of innocence (Lichtheim 124–126) to be recited after one's mortal passing and en route to one's final judgment. Kemetic people envisioned a journey to a potential future life described in the *Book of Coming Forth by Daylight* (often called the Book of the Dead). Once our mortal existence concluded, our souls would commence a journey fraught with tests and challenges that would culminate in our heart being weight against the feather of Maat to determine if one's soul could go on to exist as a maa-kher (true of voice) or whether the memory of the individual would be swallowed by oblivion (represented by the part lion, hippopotamus, and crocodile, Ammit) (Budge viii). It is worthwhile noting that in the cosmology of Kemet, there is no hell and thus the worst fate a person could encounter on the way to the future life (often called afterlife) is to be swallowed by Ammit. In other words, at this stage in history, there is no threat of eternal damnation but perhaps equally as terrifying there was the threat of being forgotten, erased from human memory.

Below is Miriam Lichtheim's translation of the 42 Declarations of Innocence, in which she too translates the word neter as God or lord, which stems from 1903 E.A. Wallis Budge's translation of the word. I take issue with that and prefer the translation of neter as "force of nature" or as I think of them: Divine principles of wisdom.

The Declarations of Innocence

Hail to you, great God, Lord of the Two Truths!
I have come to you, my Lord,
I was brought to see your beauty,
I know you, I know the names of the forty-two Gods,

Who are with you in the Hall of the Two-Truths.
Who live by warding off evildoers,
Who drink of their blood,
On that day of judging characters before Wennofer.
Lo, your name is "He-of-Two-Daughters,"
(And) "He-of-Maat's- Two Eyes."
Lo, I come before you,
Bringing Maat to you,
Having repelled evil for you.

I have not done crimes against people,
I have not mistreated cattle,
I have not sinned on the Place of Truth,
I have not known what should not be known,
I have not done any harm,
I did not begin a day by exacting more than my due,
My name did not reach the bark of the mighty ruler,
I have not blasphemed a god,
I have not robbed the poor,
I have not done what the god abhors,
I have not maligned a servant to his master.
I have not caused pain,
I have not caused tears.
I have not killed,
I have not ordered to kill,
I have not made anyone suffer.
I have not damaged the offerings in the temples,
I have not depleted the loaves of the gods,
I have nor stolen the cakes of the dead.
I have not copulated nor defiled myself.
I have not increased nor reduced the measure,
I have not diminished the arura,
I have not cheated in the fields.
I have not added to the weight of the balance,
I have not falsified the plummet of the scales.
I have not taken milk from the mouth of children,
I have not deprived cattle of their pasture.
I have not snared birds in the reeds of the gods.
I have not caught fish in their ponds,
I have not held back water in its season,
I have not dammed a flowing stream,
I have not quenched a needed fire.

I have not neglected the days of meat offerings,
I have not detained cattle belonging to the god,
I have not stopped a god in his procession.
I am pure, I am pure, I am pure, I am pure!
I am pure as is pure that heron in Hnes.
I am truly the nose of the Lord of Breath,
Who sustains all the people,
On the day of completing the Eye in On,
In the second month of winter, last day,
In the presence of the lord of this land.
I have seen the completion of the Eye n On!
No evil shall befall me in this land,
In this Hall of the Two Truths;
For I know the names of the gods in it,
The followers of the great God!

(Lichtheim 124–126)

These declarations of innocence, also known as the 42 Negative Confessions, were found in the *Papyrus of Ani,* which is dated to around 1250 BC (Scott and Swartz Dodd 333) more than a thousand years before Aristotle's writings on ethics. It is noteworthy that these declarations already contain the foundation for the ethical proto-norms that we value to this day such as the valuation of individual life, truth-telling, honesty, integrity, empathy, humility, respect for our and other people's reputation and property, etc. Furthermore, given the fact that doing the opposite of any one of these foundational values would threaten one's ability to enter a future life by weighing down our heart, it is remarkable that to this day, we speak of regrettable, sad, undesirable actions as making "our heart heavy". The fact that this is still a colloquialism speaks to the power of this ethical system to reach thousands of years forward and across thousands of miles to reach and entrench itself in a vastly different cultural context.

As mentioned at the beginning of this chapter, a Maatian approach to ethics places special emphasis on communication behaviors. Individuals were encouraged to "do" Maat and "speak" Maat. The most significant way to speak Maat was to tell the truth but there was also another aspect of Maatian communication ethics that encouraged artfulness, restraint, even silence in communication that was considered essential in order to be an excellent communicator and a speaker of "Maat". In order to show the breadth of Maatian principles in Kemetic society, below is an example from ancient Kemetic song. This excerpt is from the Triumphal Hymn of Ascension (Pyramid Text 51.1) which illustrates this point:

> Speak only that which should be spoken
> And do not speak what is not true;
> For God detests slipshod words
> I am protected! May you not misname me!
> I am the Heritor!

(Foster and Hollis 24)

Even in song form, the idea that one should communicate truthfully, that one ought to communicate thoughtfully takes center stage. As is the idea that some topics are to be broached in a specific place and time. Rather than posed as a matter of etiquette or politeness, this hymn presents these qualities, these communication practices as a matter of right and wrong. In other words, what is at stake is much more than making other people uncomfortable or damaging one's reputation. Instead, these prescriptions, speaking with care and organization among others, are presented as our human birthright as manifestations and direct scion of the sacred, of the divine. Speech was such an essential aspect of Maat that adherence to these codes of conduct would allow us to journey right into a future life as a neteru, a divinity.

The Kemetic Work of the Soul

Learning about Kemetic cosmology, one might be surprised to find that the reward for a life full of "speaking" and "doing" Maat would earn one more work in the future life. Yet, rather than the passive bliss we see represented as a heavenly existence in many of the world's religions, a Kemetic theory of the soul rewards a life of "walking the right path", a life of righteousness and truth, with an eternal but very busy existence. In order to get a more nuanced understanding of the nature and rewards of a future life, we will touch briefly on the Kemetic theory of the soul. In the Kemetic account of the future life, righteous individuals get to work the "fields of the blessed" (Weidermann 56), tending to the land, plowing with oxen, planting seeds, and doing other kinds of agricultural work from sun up to sun down. This is not exactly what most of us think of when we imagine "heaven", but as stated before, for Kemites, our cosmos offers the best and only place to dwell in any form of existence. Thus, to become "blessed" meant that Maat-observing individuals would be given the privilege to aid the continuation of life in the most direct way possible and there was no better way to do that than to tend to the life-giving soil of Kemet.

What allowed individuals to make the transition from the mortal realm to the "fields of the blessed" was the cosmic substance (spirit essence) they called *Ka* (Obenga 45). Ka itself is formless but by successfully reuniting with an individual's soul after death, it made it possible for that individual to work the fields of the blessed. *Ka* was not conceived as superior to human beings but ennobling to humans as *Ka* is the energy force that runs through the entire cosmos. However, the concept Ba represented a person's singular, individual soul (Nehusi 64). Kemites believed that mortal death entailed Ba leaving the body but if the journey through the *Duat* (underworld) and the weighing of the heart went well, the *Ba* could return to the body and reunite with the Ka to move on to another form of life of even greater importance.

The *Ka* was often represented as a bird or a hawk or even a phoenix (Carus 411). It was considered to be spiritual essence and divine. It is telling that the ntr Maat is represented as having wings and also her signature ostrich feather bore a relationship to birds. This tells us that the *Ka* was definitely associated with the practice of Maat as an ethical principle and in "doing" and "speaking" Maat Ka energy flowed through those interactions affirming and supporting life. Later on, as Philosopher Paul Carus tells us, the *Ka*: "became a human-headed hawk, a mysterious being with wings. Again, it was regarded as a spiritual essence, man's energy and will-power, obviously the product of philosophical reflection" (411).

The question of free will is the one that is of particular importance when discussing a society driven by cosmology. In a cosmological account of the nature of reality, do individuals het to choose their destiny? Are they more or less fated to become a particular type of being? How much does cosmology exert over the free will of individuals living in those societies? Maulana Karenga reminds us that while the concept of free will did not exist in Kemet as we would conceptualize it today, there is some room in the Kemetic cosmology for autonomy. He says:

> the concept of free will in ancient Egypt in no way approximates the modern individualistic concept of autonomy which suggests rights or will over and against society and/or community. For the ancient Egyptian the central moral issue is not choice but responsiveness conceptualized in the category sdm or hearing. What is of first importance to Maatian teaching is the insistence on recognition of and responsiveness to the order of things, the divine, nature and social order (253–254).

Karenga's able description problematizes the idea of free will in all of its interesting complexity. On the one hand, a cosmological account of the human experience posits a specific origin, purpose, and limitations to human beings. This was no different in ancient Greece. In the *Meno*, Socrates engages the young slave boy in the exercise of the dialectic for the purposes of assisting the young boy to recollect the knowledge Socrates was sure the boy already possessed before being born and forgetting all he knew co-existing with the forms prior to birth.

In Kemet, social order was maintained through a conservative social order that was organized hierarchically in terms of importance, prestige, and privileges. The Pharaoh as the shepherd of his people (Garcia Moreno 144) was at the top of the hierarchy. However, often omitted is that the word "pharaoh" itself means "house" and that an important part of the duties of the "house" figure was to embody, to house if you wish, the principles of Maat better than anyone else on the land. In other words, regardless of her status, the Pharaoh was still subjected to divine judgment based on the criteria set forth by Maat (Karenga 183).

Conclusion: Mythology as Allegory

Egyptologist James P Allen once described the Medu Netcher as "rich in allegory and metaphor but relatively poor in vocabulary" (13). In this chapter, we took the necessary steps to understand the communicative dimensions of Maat in their proper cosmological and spiritual contexts. Throughout our discussion, we have pointed to the groundedness of the Kemetic spiritual view; we have explored the differences between contemporary religious practices and an ancient spirituality rooted in place and community among all cosmic substances.

A possibility we are yet to explore is that perhaps Kemites were as, or even more, attracted to allegory and poetry that many societies are today. There is, after all, evidence of a Kemetic predilection for symbolism and metaphor. As one Egyptologist notes:

> A typical Egyptian view of the soul is a description of the sentiment that throbs in our breast-that part of the body that lies between the arms and finds a vivid expression in the use of our hands. It is called ka and is pictured in hieroglyphics by two out stretched arms, which is commonly translated "double," for it is supposed to be the ethereal shape of the man and represents the personality as a kind of astral body, which is supposed to be in

possession of all attributes of the man to whom it belongs. The translation "double" is in so far justified as the monuments actually represent the ka as a second and an additional figure, which, at certain times and certain places, is deemed necessary to add to the representation of a man. We see, for instance, the picture of a new-born prince in which his double, his idealized self, is represented right behind him, bearing a special name, the so-called fca-name of the future king (Carus 420).

Speculation and theories about the meaning and significance about the doppelganger, shadowy figures portrayed in Kemetic art abound but if we rid ourselves of the prejudice than an ancient culture which is incapable of communicating in any way other than literally, there might be room to interpret these shadowy figures as our "potentiality". As mentioned earlier in the chapter, the neteru are there to remind us constantly of the things we ought to value and the power that can be available to us if we "walk the right path" and lead righteous lives. Perhaps, these shadows are a poetic, visual representation of the other aspects of ourselves that need to be brought to light. An ethics, after all, has to be demanding by definition in order to fulfill its aspirational function.

I don't think it is far-fetched to conceive of a culture so accomplished in many other regards to also exhibit major accomplishments in communication arts: In poetry, in song, in literature, and other forms of indirect communication and I'm not alone in this idea. Asante and Mazama present a clear picture of African societies, including Kemet, in which proverbs, teachings, and even epistemological ideas were presented in allegorical form (133, 137, 244) to, among other things, enhance reception and recall. Anthropologist and African Studies scholar Dona Richards goes even further by claiming that: "The predominant mode, in terms of metaphysical explanation, is symbolism. It is here that so many difficulties and distortions arise when African culture is approached and described by the literal-minded" (219).

Is the story of Maat and her role in the pantheon of neteru also an allegory? Are these creation stories poetic ways of teaching and learning about morality? Have Egyptologists underestimated this civilization by interpreting as literal their stories and artistic expression in relation to ethics?

Today, scholars have come to understand that one of the principal features of the mythology corpus is its role as morality tales for many generations (Gotlieb 478). If we agree this is the nature of myth, then such elaborate cosmology should also alert us of other cultural behaviors and practices that at some level have already been ratified and codified into parabolic narratives that function as pedagogies of morality. In this case, Maatian ethics is that other foci of cultural behaviors and practices codified as good, as right, and as desirable. In the next chapter, we will learn more about the various ways in which Kemites sought to enlighten their intellect and pursue their full potential through scientific study and how this educational quest is connected to doing and speaking Maat.

Bibliography

Allen, James P. *The Ancient Egyptian Pyramid Texts.* vol. 38. Society of Biblical Literature Press, 2015.

Aristotle. *Metaphysics.* Translated by C. D. C. Reeve. Hackett Publishing Company, 2016.

Asante, Molefi K. and Ama Mazama. *Encyclopedia of African Religion.* SAGE, 2008.

Budge, E. A. Wallis Sir. *Tutankhamen, Amenism, Atenism, and Egyptian Monotheism: With Hieroglyphic Texts of Hymns to Amen and Aten,* translations, and illustrations. M. Hopkinson and company Ltd., 1923.

Carus, Paul. "Conception of the Soul and the Belief in Resurrection among the Egyptians (Illustrated)." *Monist,* vol. 15, no. 3, 1905, pp. 409–428.

Case, Shirley J. "Christianity and the Mystery Religions." *The Biblical World,* vol. 43, no. 1, 1914, pp. 3–16.

Colla, Elliott. *Conflicted Antiquities: Egyptology, Egyptomania, Egyptian Modernity.* Duke University Press, 2007.

Cunningham, Michael R. "Weather, Mood, and Helping Behavior: Quasi Experiments with the Sunshine Samaritan." *Journal of Personality and Social Psychology,* vol. 37, no. 11, 1979, pp. 1947–1956.

Foster, John L. and Susan T. Hollis. *Hymns, Prayers, and Songs: An Anthology of Ancient Egyptian Lyric Poetry.* vol. 8. Scholars Press, 1995.

Gottlieb, Alma. "Dog: Ally or Traitor? Mythology, Cosmology, and Society among the Beng of Ivory Coast." *American Ethnologist,* vol. 13, no. 3, 1986, pp. 477–488.

Hasan, Fekri A. and Shelley J. Smith. "Soul Birds and Heavenly Cows: Transforming Gender in Predynastic Egypt." *In Pursuit of Gender: Worldwide Archaeological Approaches.* vol. 1, edited by Sarah M. Nelson and Myriam Rosen-Ayalon. AltaMira Press, 2002, pp. 43–65.

Karenga, Maulana. *Maat, the Moral Ideal in Ancient Egypt (African Studies).* Taylor and Francis, 2004.

Klimkeit, Hans-J. "Spatial Orientation in Mythical Thinking as Exemplified in Ancient Egypt: Considerations toward a Geography of Religions." *History of Religions,* vol. 14, no. 4, 1975, pp. 266–281.

Lampert, Jay, "Hegel and Ancient Egypt: History and Becoming." *International Philosophical Quarterly,* vol. 35, no. 1, 1995, pp. 43–58.

Lant, Antonia. "The Curse of the Pharaoh, or How Cinema Contracted Egyptomania." *October*, vol. 59, 1992, pp. 87–112.

Laukhuf, G. and H. Werner. "Spirituality: The Missing Link." *The Journal of Neuroscience Nursing: Journal of the American Association of Neuroscience Nurses*, vol. 30, no. 1, 1998, pp. 60–67.

Lichtheim, Miriam. *Ancient Egyptian Literature Volume 1: The Old and Middle Kingdom*. University of California Press, 1975.

Lynn, Marvin. "Toward a Critical Race Pedagogy: A Research Note." *Urban Education*, vol. 33, no. 5, January 1999, pp. 606–626.

Malunga, Chiku. "Identifying and Understanding African Norms and Values that Support Endogenous Development in Africa" *Endogenous Development: Naïve Romanticism or Practical Route to Sustainable African Development (Development in Practice Books Series)*, edited by Chiku Malunga and Susan H. Holcombe. Routledge, 2014, pp. 9–22.

Marumo, Phemelo O. and Mompati V. Chakale. "Understanding African Philosophy and African Spirituality: Challenges and Prospects." *Gender & Behaviour*, vol. 16, no. 2, 2018, pp. 11695–11704.

Moreno Garcia, Juan C. *The State in Ancient Egypt: Power, Challenges and Dynamics*. Bloomsbury Academic, 2019.

Nehusi, Kimani S. K. "The Construction of the Person and Personality in Africa." *Regenerating Africa: Bringing African Solutions to African Problems*, edited by Mammo Muchie et al. Africa Institute of South Africa, 2017, pp. 61–76.

Ntwadumela, K. "A Tribute to Dr. Yosef Ben Jochannan: The Black Wombman as God." *The Journal of Pan African Studies*, vol. 8, no. 10, 2016, pp. 97–123.

Obenga, Théophile. "Egypt: Ancient History of African Philosophy." *A Companion to African Philosophy*. vol. 28, edited by Kwasi Wiredu et al. Blackwell Publishing, 2006.

Park, Hyungji. ""Going to Wake up Egypt": Exhibiting Empire in Edwin Drood." *Victorian Literature and Culture*, vol. 30, no. 2, 2002, pp. 529–550.

Quirke, Stephen. *Exploring Religion in Ancient Egypt*. John Wiley & Sons Inc, 2015.

Richards, Dona. "The Nyama of the Blacksmith: The Metaphysical Significance of Metallurgy in Africa." *Journal of Black Studies*, vol. 12, no. 2, 1981, pp. 218–238.

Robertson, Robin. "As above, so below." *Psychological Perspectives*, vol. 57, no. 4, 2014, pp. 403–425.

Robins, Gay. "The Names of Hatshepsut as King." *The Journal of Egyptian Archaeology*, vol. 85, no. 1, 1999, pp. 103–112.

Scott, David A. and Lynn Swartz Dodd. "Examination, Conservation and Analysis of a Gilded Egyptian Bronze Osiris." *Journal of Cultural Heritage*, vol. 3, no. 4, 2002, pp. 333–345.

Shaw, Gary. "A Matter of Life and Death: The Popular View of Ancient Egypt as a Culture Obsessed with Death Is Challenged by Two Exhibitions, Which Argue that Funerary Objects Were as Important in Daily Life as They Were in the Afterlife." *Apollo*, February 1, 2016, p. 62.

Teeter, Emily. "Egypt." *The Cambridge Companion to Ancient Mediterranean Religions*, edited by Barbette S Spaeth. Cambridge University Press, 2013, pp. 13–33.

Ward, William A. "A Unique Beset Figurine." *Orientalia*, vol. 41, no. 2, 1972, pp. 149–159.

West, C. S'thembile. "The Goddess Auset: An Ancient Egyptian Spiritual Framework." *Goddesses in World Culture*, edited by Patricia Monaghan. Prager, 2010, pp. 237–248.

Wiedemann, Alfred. *The Ancient Egyptian Doctrine of the Immortality of the Soul.* G. P. Putnam, 1895.

Wilkinson, Charles K. *Egyptian Wall Paintings: Metropolitan Museum of Art's Collection of Facsimiles.* Metropolitan Museum of Art, 1983.

3 Scientific Communication and the Divine

As perhaps, the first thinker to make human communication as such a philosophical problem (Peters 128), Søren Kierkegaard focused his attention not just on the content of our conversations about ethics, but its form. He approached the question of what is to communicate through a fourfold division: (1) The object of inquiry; (2) the communicator, (3) the recipient, and (4) the communication (Roberts 51). Kierkegaard saw his philosophical task as exploring "suitable" ways in which ethical and ethico-religious truths ought to be communicated in order to create impetus for individual and collective action in accordance to ethical principles. He concluded that regardless of the particularities of any ethical or ethico-religious belief system, the presentation, reflection, and internalization of the ideas put forth by individuals engaged in these conversations encompass a distinct, multi-parted, and philosophically interesting communicative process.

Kierkegaard saw discourse on ethics as a multi-layered process that requires both objective and subjective components. Additionally, it requires corresponding direct (objective) and indirect (subjective) types of communication for its proper expression and practice. In other words, for Kierkegaard, ethics is not about the communication of (ethical) knowledge alone, but about the communication of (individual) ability, which, by his account, requires both a deeper (more aesthetic) understanding of ourselves and a type of communication that doesn't rely solely on scientific, empirical, or objective truths, but also compels us to examine our very consciousness to develop a sense of purpose for our individual relation to the world. Ethical communication then, for Kierkegaard, should stimulate interlocutors to reflect inwardly to examine their own, subjective, commitments and abilities or, as he puts it, the individual's capacity to "be by doing" (Roberts 56).

In examining the dialectical relationship between scientific and spiritual knowledge as it relates to ancient Kemetic Maatian principles, I find Kiekegaard's dual framework of direct and indirect communication to be the closest and most helpful to tease out the Maatian principles involved in this way of life. At the heart of Kierkegaard's concerns lies an intense preoccupation with bridging the gap between "knowing" what the "right thing to do is", like say the principles of Maat, and the gap that is often present between "possessing" this "knowledge" and practicing this knowledge in our daily ethical interactions. He wanted to explore what "possessing" this information really means: Can moral principles be transferred or exchanged with others? Are ethical principles only tools through which individuals negotiate relationships with others? Or are they the means by which individuals negotiate relationships with themselves?

Maat, of course, as a moral idea must not be imposed on individuals by means of coercion or threat; it must be freely chosen. Ethical approaches manifest themselves from within, but if that is the case, what made Maat and what makes any other ethical principle binding? How can the approach be taught or popularized if force or law is not leveraged as incentives? Kierkegaard understood that though deeply interrelated, thought and being are not equivalent (Roberts 3). He utilizes this distinction as the point of departure for his analysis partly because he wants to grant Immanuel Kant the distinction between the phenomenal and the noumenal and therefore wants to conceptually limit human knowledge to the objects of our experience.

Therefore, while he wants to maintain that "being" has a dimension that is unknowable to us, since we cannot have a God's eye point of view, the objects of our experience are definitely knowable and understandable to human beings through our ability to think. Implied here is the inference that because the objects of our experience are knowable, then we ought to shift our focus and inquiry to them. In the case of Kemet, they too created an epistemological category for the unknown although in their cosmology *Nun*, the primeval waters out of which the world arose are also associated with chaos. As Egyptologist Jean Yoyotte notes:

> Ptah conceiving in 'his heart which is Horus' and creating via 'his tongue which is Thoth'; *Sia*, 'knowledge', and *Hu*, 'order', major attributes of the sun; the four Souls which are *Ra* (fire), *Shu* (air), *Geb* (earth) and *Osiris* (water); the unknowable and infinite God who is 'the sky, the earth, the Nun, and everything that lies between them'; and so on (126).

This is consistent with a Maatian ethic as the ability to maintain and propagate order is deemed as the height of that which is wise, desirable, and divine in the Kemetic value system. Therefore, by the same logic, chaos and disorder would be categorized as that which is unknowable and foreclosed. In this respect, while there is an acknowledgment of a limitation to human knowledge, the unknown appears to have a pejorative affiliation with *Nun*. This is not the case for either Kant or Kierkegaard for whom the unknowable and the noumenal were necessarily and unavoidably part of the human experience. This makes sense given the fact that both Kant and Kierkegaard are operating with a concept of God that is separate, external from human beings and nature, while Kemites worked with a different, consubstantial model in which no such separation was theorized.

Still, in terms of the distinction between thought and being, the differentiation does apply to Kemetic thought and it is of particular relevance to the practice of ethical communication. Though for Kierkegaard, being and thought are not the same thing, thinking is, in fact, a kind of activity (of being) that plays a crucial role in our interpretation of experience, our identity, and our commitments as an object of inquiry. The word interpretation is crucial here as human beings lack certainty with respect to what "things in themselves" are. However, this limitation should not be an obstacle to our ethical endeavors. Instead, the limitation of our knowledge should remind us of the fact that as individuals living and interacting with others in this world, we are in a constant conversation not just with other human moral agents but with our own thoughts.

Herein lies a key insight into the contrast between a European view on the limitation of human knowledge and a Kemetic Maatian view. For Kierkegaard and Kant, the fact that there are limits to what humans can know should have a humbling effect on our perception of our own knowledge compared to, say, God. This realization, this fact, if you'd like, should work to temper human hubris, which indeed is a moral disposition. However, when we contrast this concept of the unknown to that of Kemet, we do see a general admission of the limits of human knowledge, yes, but we also see the presence of the neteru as a constant reminder of what we can aspire to be. We see examples of what a fully actualized cosmic substance looks like and acts like. In this context, the neteru inspire and motivate to both behave ethically and diversify our sphere of knowledge as there are not one but 42 aspects of the nature and the sacred to study, know, and emulate.

Monologue as Internal Dialogue

Maatian ethics also places emphasis on the conversations that moral agents have with themselves. As part of a larger, older spiritual reflection practice (Malamba 217), speaking the 42 laws of Maat, in either the affirmative or their negative articulation, is an example of a conversation that the moral agent carries out internally, at least before they get to the day of final judgment. Furthermore, they exemplify the type of thinking activity that is also a form of reflective doing (being). This view of human communication encompasses not just dialogic encounters with others but an essential and necessary form of communication with ourselves (monologue). This process is intimately connected to the question of why it is that while most of us seem to be cognizant of how we are expected to treat each other, through laws, codes of conduct, or even common sense, some of us remain apathetic or, worse, defiant of these expectations and behavioral norms. Similarly, the Kierkegaardian project is largely dedicated to shedding light into this most private and internal type of communication which he calls "maieutic" following Socrates' name for his Socratic method.

According to Don Roberts in his *Kierkegaard's Ethical and Ethico-Religious Dialectic of Communication*, maieutic communication includes monologue and other types of indirect communication such as anonymity, humor, metaphor, story-telling, etc. Such techniques of speech are put at the service of teaching morality without issuing commands or unduly influencing the listener. Roberts notes maieutic communication:

> must neither be an attempt to transmit information nor an attempt to effect a meeting of the minds in regard to some proposition, but must rather be an attempt to stimulate the listener into self-activity as is necessary in order that he may come to know the truth by himself (13).

The fact that such an approach places a heavy emphasis on listening as well as in stimulating individual reflection is not only a sign of the kind of reflective praxis associated with ethical approaches, but specifically of a Maatian approach. For instance, the concept of self-activity highlights the many ways in which Maatian ethics place accountability and judgment squarely on the shoulders of each individual moral agent. Self-activity can very much fit the framework of spiritual activity that enables solitary reflection, self-evaluation, and the affirmation of one's moral goals, as well of the means by which to achieve these goals.

Maat as the Model of Maieutic Communication

Also known as Ma'at or Mayet (Premnath 324), the moral ideal of Maat as well as its anthropomorphized woman figure with her wings outstretched work to propagate the ethical ends of the Kemetic people of harmony, balance, justice, etc. For this reason, a Maatian ethical approach is listening-centered and is understood as an essential means by which individuals can achieve their full spiritual potential. However, achieving one's full potential also requires evaluation and reflection along the way. It requires one to devise a series of goals and the means to achieve those goals in order to stay the course toward full actualization. In this sense, Maat is also at the service of spirituality.

In the Greek context, the Socratic method is also seen as a means by which individuals (philosophers) can aid others to recollect, and in doing so, become a better, fully actualized version of themselves. The Socratic approach is well known for its questioning approach. In framing the role of the philosopher as a questioner, the end goal of this form of communicative interaction is to allow the interlocutor to recall, to come to the realization, that the answers to the questions being asked are somewhere deep within themselves. To be sure, this is an idiosyncratic way of teaching. The teacher must take every precaution not to exert undue influence on the student, making that student dependent on the teacher for answers.

There are good reasons why the question and answer format is presented by Socrates as the most effective way to practice teaching: (1) It allows students to search within themselves to provide answers to the questions posed; (2) in the Greek cosmology, by remembering the answers to the questions posed by the philosopher, the student reconnects with their true self or the self that was forgotten due to the trauma of transitioning at birth from the world of the ideal forms to a mortal existence; and (3) finally, from the perspective of the philosopher, this approach helps to keep the teacher engaged with the community and keeps their ears open to new interpretations, new answers, and new facts that help the philosopher continue to develop in depth and breadth of knowledge.

Interestingly, other scholars have pointed out important parallels between Socrates approach to communication ethics, moral pedagogy, and Maat. Ethicist Maulana Karenga, for example, has outlined the similarities in the fundamental assumptions each approach makes about the nature of vice or immorality. He states:

> Socratic moral theory on knowledge and virtue has essentially four basic interlocking contentions, most of which find parallels in

Maatian moral theory. These are: 1) no one voluntarily chooses to do what one considers to be bad; 2) the possession and practice of virtue requires knowledge and can be equated with it; 3) the object of moral knowledge is the Good; and thus 4) wrongdoing reflects and is rooted in ignorance of the Good. Although Maatian ethics does not specifically state that knowledge is virtue and ignorance is vice, as does Socratic moral theory, it does assert that knowledge is virtuous and corrective and ignorance is vicious and corruptive. Therefore, the assumption is made, as in Socratic theory, that if one knows the Good, s/he will choose and seek it (245–246).

Etymologically, there appears to be an even closer link between the Socratic Maieutic approach to communication and Maat. The term Maieutic is derived from the Greek word for midwife (*maieutikos*) (Hanke 459). The Greek meaning of the term has thus far been taken to be descriptive of the approach as the Socratic method is seen as a way to assist in the (re)birth of the student as she is able to recollect information that her old self accessed in the world of the forms (pre-existence). Additionally, the term describes the birth of philosophical knowledge. As one scholar explains:

> One approach to the analysis of Platonic communication theory is based on the metaphoric description of Socrates' method of midwifery (maieutic). The dyad of the pregnant one and the midwife corresponds to the production of philosophic knowledge by argumentative dialogue, in and by which a thought is 'brought out'. Being an exchange and interpretation of signs, Platonic dialectic can be regarded as a semiotic process. Furthermore, being dyadic and mutual dialectic is a joint activity and (philosophic) communication (Hanke 459).

Thus, the term maieutic is symbolic of the kind of intellectual labor that produces and introduces new ideas, new concepts, and perspectives into public discourse. The role of the teacher/philosopher is to, in a sense, coax this knowledge out of the students while allowing the student to do the intense work of pushing those ideas outside of themselves. Which, to me, begs the question: If Socrates intended to highlight the collaborative aspect of knowledge creation or the role of the philosopher as a facilitator or mediator in producing wisdom, why not use the analogy of a bridge? Or the blooming of a flower? Or jumping over a fence? Or a priest? Or rowing a boat? Why something so straightforwardly womanly? One clue lies in the etymology of the term for midwife.

Maia as in "Maieutria" (the use of hands in a birth delivery) or midwife comes from Greek mythology where Maia is said to have birthed none other than Hermes (Greene 344). Hermes is now known to be the Greek version of Thoth. Of this fact, there is little doubt. As one scholar puts it, "Scholars have generally accepted that the earliest reference to the equation of Thoth and Hermes Trismegistus occurs in a Greek papyrus document of the end of the century BC." (Skeat & Turner 208).

So far, we have established that the Socratic term "maieutic" is a reference to the Greek term for midwife (maieutikos) and that the root word for that is "maia". Maia, in the Greek mythological pantheon, is the mother of Hermes, who scholars now know is the Greek name for "Thoth", the Kemetic ntr of wisdom, knowledge, logic, and reason. But is there more to this coincidence? Is it possible that Socrates was making a direct reference to Maat when we termed his teaching philosophy "Maieutic"?

Per Greek mythology, Maia, the oldest of the seven Pleiades, and Zeus were the parents of Hermes. Maia is also known as Theogonies and commonly referred to as "Gaia" the earth mother" (Colavito 10, 45). According to Kenetic cosmology, Maat was the daughter of the ntr *Ra* (*Re*) and the ntr *Hathor* (Sekhmet) (Fletcher 68), while in Greek mythology Maia was the daughter of Atlas and Pleione the Oceanid (Akulov 15). Atlas is associated with Shu (the ntr of air) and Pleione is not broadly associated with any Kemetic ntr.

While the ancestry does not align, it is an interesting fact that Maia was in Greek mythology the mother of Hermes (Djehuti, Thoth) who is well known for his association with not just knowledge (Thoth, "Thought") and wisdom but specifically Kemetic knowledge and wisdom. That symbolic parallel seems meaningful and intentional. Additionally, our discussion in Chapter 2 touched on the Medu Netcher symbol for Ka, the two out stretched arms, mimicking, if you'd like, to a manuduction (birth delivery by hand).

A third interesting parallel between Maia and Maat is the fact that in Greek mythology, Maia (along with her six sisters) are transformed into doves by Zeus (Rigoglioso 154) which is reminiscent of Maat's iconic winged-woman figure. Finally, while proving a definitive connection between the Maieutic method and Maat is better left to linguists and classicists, it is worth noting that the pronunciation of Maieutic and the pronunciation of Mayet come shockingly close. Try it:

"M a i e u t - i c" vs. "M a y e t - i c"

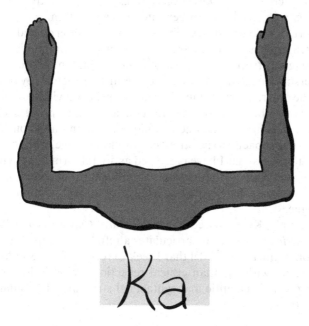

Figure 3.1 Arms in posture. Gardiner Code D28.

Kierkegaard identifies a direct form of communication that is objective in character and that he claims is central to sharing, and indeed building scientific knowledge from which everyone can and should benefit. Direct (external, objective) communication is presented as more effective for the purposes of bringing people together in conversation about empirical facts, while indirect (internal) communication accounts for the other integral part of this ethical education (and life) that, in his view, requires reflection if only to increase the individual's self-knowledge by objectifying her own thought process and by extension one's ability to relate to authenticity and honesty with others in her community.

The difference, then, is that while both indirect (maieutic artistry) communication (Kellenberger 154) and direct communication should be situated in the world of experience, direct communication is better utilized as a medium to build consensus based on more concrete, quantifiable, or scientific information. Ultimately, both types of communication are integral to self-actualization as indirect communication is crucial in allowing the individual to harmonize with herself, while direct communication allows us to learn, evaluate, and share factual, objective data. If either one of these communicative abilities is stifled, a gap will surface between intention and action.

In short, Kierkegaard describes at least two interrelated, yet differentiated processes necessary for the proper communication of ethico-religious truths. One process, interpersonal in nature, involves the understanding and legitimization of ethical principles by individuals and the other, internal, equally influences the individual to practice or disregard ethical principles through a dialogic process of self-discovery. In other words, one technique of communication is more related to our capacity to acquire objective knowledge, while the other involves a process that I have presented as a vital form of activity (being), reflective activity, that requires a unique form of communication that is sometimes disregarded as a form of ethical communication: Monologue.

For example, Kwasi Wiredu, in his *Cultural Universals and Particulars: An African Perspective*, articulates a refutation to the type of internal communication model that I seek to recover from Kierkegaard and associate with Maatian ethical reflection. Wiredu does this by appealing to the semantic and syntactical structure of language. He states that:

> worse, no one could converse with him or herself, for any kind of conversation at all presupposes syntactical and semantic rules which, if they are available to any one individual, would argue the existence of regularities of which others, too could, in principle, avail themselves for the purpose of communication (14)

One possible answer to Wiredu's refutation may be to put pressure on his definition of "conversation". The reader might ask, could it be that the type of exchange Wiredu has in mind is significantly different from the type of internal communication that Kiekegaard seems to be describing to make this an equivocation? One might even question whether the type of framework necessary for individuals to "converse" with themselves would necessarily require the exclusion of language. Why should a monologic encounter be considered impossible only because it might be (or is) mediated by language, which is intrinsically a product of our interactions with others?

It seems to me that there is a sense in which a conversation or a dialectic encounter boils down to the elucidation of two or more differing points of view and that as long as an individual is exploring these points of view with great vigor while trying to decide which one is more persuasive or the "right" ethical course (path) of action, then a significant and, in fact, multi-voiced intellectual exchange is taking place. This might seem counter-intuitive at first, but in light of modern psychoanalytic, and broader critical perspectives, we should at least question what we mean by persons (as in the case of a present and physical interlocutor) and the extent to which, as Judith Butler argues, we play an active role in constructing the "other" that we are positioning as our dialogic partner. This is, of course, not to say that since the other doesn't exist except in our heads, then interaction with others is a fantasy, but maybe one ought to consider the possibility that more than a "concrete" physical agent, when it comes to ethical conversations and reasoning, we might be engaging more than an interlocutor, but a position, a perspective that while grounded in particularity and historical reality, it may lack the physical component that seems to be required by many contemporary ethical theories which are based on dialogue.

Direct Communication as Divine

As early as 1841, when Kierkegaard completes his *Concluding Unscientific Postscript to Philosophical Fragments*, he is worried that moral theories have become too preoccupied with "what" it is to be highlighted as a characteristic of the human condition or as the basis for theories of just distribution. He is concerned that the theories of his time overlook critical issues such as what communication is (especially ethical communication) and how it is deployed by individuals. One aspect of Maat that propels individuals to engage in a direct, meaningful, dialogic practice is the pursuit of scientific knowledge.

Sacred science is a reflection of the sacredness that Kemites attributed to all cosmic substances. Scientific exploration played a crucial part in the spiritual actualization of each human being. While Kemet is wildly recognized as home to some of the wonders of the world and as having achieved technological and scientific feats that defy explanation even by today's standards, little attention is paid to the type of value system that fostered such intellectual curiosity and rewarded such scientific innovation One scholar explains with great clarity the connection between the study of the laws and patterns found in the physical world applied to the science of metallurgy, to spirituality in the Kemetic sense of the term. She says:

> Because spiritual realities or cosmic truths manifest, reflect, or reveal themselves in corresponding material being, the 223 sacred secrets of the universe can be extracted from earthly substances. Thus

working with metals becomes an attempt to extract knowledge from a human activity; it becomes a spiritual and learning process. In order to make this clearer, we must understand that the African sense of cosmic order is so complete, detailed, and profound that the ultimate is intimately related to the most minute in this conception. Every realm is a reflection of every other. The way the culture is organized reflects the relationship between the universe, the society, and the person. The social units are not arbitrary but are seen to have evolved from the sacred order. The explanation of the cosmic order, cosmology, generates a theory of reality and being: an ontology. It is a series of circular relationships (Richards 223).

Although today we may think of the term sacred science in terms of deep ecology, indigenous healing practices, or even the romantic naturalism of Ralph Waldo Emerson, Kemetic society saw benefits in combining science and spirituality (Martin 160). As one of the first settled civilizations in history, "Egypt was settled by refugees from the deserts of the eastern Sahara and the southern Levant, fleeing from mid-Holocene droughts, and became a melting pot of indigenous Nilotes and desert herders, part-time cultivators, and hunters." (Hassan 135). As a result, they evolved from pastoralist and agrarian societies who were keen observers of patterns in the physical universe, including earth and sky (Browder 74). This knowledge afforded them superior predictive abilities, and played an important role in the development of commerce, the sustenance of an exploding population, and in challenging themselves to secure the memory of their ancestors and would-be ancestors with ever more elaborate, majestic buildings and other commemorations of their lives and legacies.

To Kemites, science was seen as spirituality, the sort of objects of experience Kierkegaard was keen for human beings to devote time and effort to decipher. One of the most important imperatives in the Pan-African worldview is to "know thyself" (Mungwini 5) and while most of us would take this to mean that we ought to investigate our own thoughts, beliefs, and aspirations, in Kemet, this also meant to get to know the physical world, the natural world and all of its inhabitants, human or not. This is a time and a place in history when Cartesian dualism is far ahead in the future and in its place. There is a different telos: To achieve complete oneness with the universal consciousness (*Ka*).

Following this logic, this is a time in which it also would've been a person to truly know themselves without conducting the necessary study and inquiry into the human place in nature, After all, part of our humanity is that we are embedded and have a relationship of interdependence; thus realistically, a person, a human creature, cannot possibly know herself in any meaningful or thorough capacity unless she has sought knowledge about her material conditions. It is with this mindset that Kemet embarked on a societal and moral quest for knowledge of themselves, which, of course, included nature. This excerpt for ancient Kemetic literature titled "The Instruction of Little Pepi on His Way to School" illustrates the widespread belief in the importance of educating oneself in a wide range of subjects:

> 'Now, it is good to study many things that you may learn the wisdom of great men. Thus you can help to educate the children of the people while you walk according to the wise man's footsteps. The scribe is seen as listening and obeying, and the listening develops into satisfaction. Hold fast the words which hearken to these things (Foster 42).

The School of Alexandria

Although Kemet has no shortage of great minds and innovators in every scientific endeavor of their time including philosopher, physician, Imhotep and criminally understudied mathematician and philosopher Hypatia, perhaps no other image captures the vitality of Kemet as a center for intellectual and scientific inquiry more than the School of Alexandria. Truly, prior to that many philosophers from Egypt and surrounding areas, including Greece, came to study at the Egyptian Mystery Schools (Pythagoras of Samos, perhaps being the best-known alumni), but the School of Alexandria and its immeasurable contributions to medicine and Christian ethics seem to have captured the world's imagination in a unique way. As one scholar notes its unsurpassed legacy in human medicine:

> The great tradition of ancient Egyptian medicine was maintained for millennia. The *Edwin Smith papyrus* and the *Ebers papyrus* both speak of exquisite knowledge and understanding. The *Edwin Smith papyrus* is the earliest known medical document, written around 1600 BCE, but is thought to be based on material from as early as 3000 BCE. It is an ancient textbook on trauma surgery. It mentions trepanation. It describes anatomical observations and the examination, diagnosis, treatment, and prognosis of numerous injuries in exquisite detail. It gives the first descriptions of the cranial sutures, the meninges, the external surface of the brain, the cerebrospinal fluid, and the intracranial pulsations (Serageldin 1).

The imperative of knowing oneself would have obvious implications for the study of the human body and the various conditions that impact its wellbeing but such a zeitgeist is bound to manifest itself through the communication practices and ethics of people. And it did. Sophists received special training on rhetoric and speech in its campus. Christian ethicists like Clementine infused the teachings with a morality that became its flagship pedagogy. Throughout its history, the School evolved and cross-pollinated with the Greco-Roman dominance of its time. As one scholar laments:

> The Alexandrian school, though located on Egyptian soil, was essentially Greek in its personnel and habits of thought. About 80 B.C. Egypt came under Roman domination. After this time the character of the school gradually suffered a change. The earlier scholars had devoted themselves to science and literature, while in later times their main interest was in what we would now call philosophy. Yet while it lasted (until the fifth century A.D.) the school of Alexandria included some great names: Euclid (about 300 B.C.); Apollonius of Perga (200 B.C.), the author of a treatise on conic sections; Eratosthenes (230 B.C.), who made the first measurement of the circumference of the earth; and Hipparchus (160-125 B.C.), who found the epicyclical theory of the heavens, later known as the Ptolem (Heyl 141–1420).

The school's contribution to the study of rhetoric as a form of sacred speech can hardly be underestimated. Today, a proper scholarly study of the classical rhetorical cannon must include the achievements of the School of Alexandria. As noted in one such syllabus: "It is there that we find the oldest center of sacred rhetoric. After considering the school of Antioch and the African rhetoricians" (Gythiel 6). Thus, while sacred rhetoric is something that was studied and practiced by ancient African people long before the School of Alexandria came along, its contribution to the development of that field elevated the indigenous study of rhetoric through its immersion and affiliation with many other scientific pursuits at the School of Alexandria.

Conclusion: Morality as Technology

As part of the Maatian approach to ethics, ethical knowledge encompasses much more than knowledge about or of ethics. It encompasses the pursuit of knowledge of the physical manifestations of the divine in which one is embedded. Part of a virtuous character, then, is to pursue this knowledge through the use of one's own abilities for thought and communication.

In this context, to live according to one's own nature was meant in the broad sense of the term that included both spiritual and physical dimensions. In this short chapter, I've tried to highlight the moral impetus of the insatiable Kemetic pursuit of scientific knowledge while framing it in the context of its Maatian ethical imperative. Perhaps Africana Studies giant Théophile Obenga puts it best when he says:

> Egyptian thought made the greatest achievements in the fields of philosophy (wisdom) and science, i.e. astronomy, medicine, architecture. But spirituality ("religion") and morals were not neglected. In all these fields the Egyptians sought truth and certainty through rational inquiry (Obenga 44)

Given the proclivities and reach of the Maatian ethic approach to spiritual science, it is no wonder why someone of the depth and erudition of Socrates would have potentially been inspired, motivated, and dedicated to the pedagogical wisdom contained in Maat. Is it a stretch to concede that "Maieutic artistry" was a reflection of "Mayetic or Maatian" ethical principles? More work needs to be done to confirm or disconfirm my theory but what is clear to me is that it is no longer permissible to claim knowledge or expertise in the Socratic method without an earnest effort to acquaint oneself from what appears to be its likely source of inspiration.

If what I suspect is true and Socrates was indeed referencing the Maatian (and as shown Kierkegaardian) supposition that there is a communicative science to doing the right thing and that spirituality is unseen science, then understanding the Kemetic spiritual, ethical, cultural, and historical contexts in which these ideas and techniques are formed can no longer be considered a choice but an intellectual and ethical imperative.

Bibliography

Akulov, Alexander. "A Minoan deity from London Medicine Papyrus." *Cultural Anthropology and Ethnosemiotics*, vol. 3, no. 2, 2017, pp. 13–17.

Bracci, Sharon L. and Clifford Christians. *Moral Engagement in Public Life: Theorists for Contemporary Ethics*. Peter Land Publishing, 2002.

Browder, Anthony T. *Nile Valley Contributions to Civilization: Exploding the Myths*. vol. 1. The Institute of Karmic. Guidance, 1992.

Colavito, Maria M. *The New Theogony: Mythology for the Real World*. State University of New York Press, 1992.

Fletcher, Joann. *The Egyptian Book of Living and Dying*. Thorsons, 2002.

Foster, John L. *Ancient Egyptian Literature: An Anthology*. University of Texas Press, 2001.

Greene, Elizabeth S. "Revising Illegitimacy: The Use of Epithets in the Homeric Hymn to Hermes." *The Classical Quarterly*, vol. 55, no. 2, 2005, pp. 343–349.

Gythiel, Anthony P. "Classical Rhetoric: Pagan and Christian Syllabus." *Rhetoric Society Quarterly*, vol. 10, no. 1, 1980, pp. 2–8.

Hanke, Michael. "Socratic Pragmatics: Maieutic Dialogues." *Journal of Pragmatics*, vol. 14, no. 3, 1990, pp. 459–465.

Hassan, Fekri A. "The Predynastic of Egypt." *Journal of World Prehistory*, vol. 2, no. 2, 1988, pp. 135–185.

Heyl, Paul R. "The Genealogical Tree of Modern Science." *American Scientist*, vol. 32, no. 2, 1944, pp. 135–144.

Karenga, Maulana. *Maat, the Moral Ideal in Ancient Egypt (African Studies)*. Taylor and Francis, 2004.

Kellenberger, J. "Kierkegaard, Indirect Communication, and Religious Truth." *International Journal for Philosophy of Religion*, vol. 16, no. 2, 1984, pp. 153–160.

Kierkegaard, SØren. *Concluding Unscientific Postscript to Philosophical Fragments*. vol. 2. Edited and translated by Howard V. Hong and Edna H. Hong. Princeton University Press, 1992.

Malamba, Theodore M. "African Heritage in the Global Encounter of Cultures." *Communication across Cultures: The Hermeneutics of Cultures and Religions in a Global Age*, edited by Chibueze C. Udeani, et al. Council for Research and Values in Philosophy, 2008, pp. 213–238.

Mosha, R. S. *The Heartbeat of Indigenous Africa: A Study of the Chagga Educational System*. vol. 1442. Garland Publishing Inc., 2000.

Mungwini, Pascah. "African Known Thyself": Epistemic Justice and the Quest for Liberative Knowledge." *International Journal of African Renaissance Studies - Multi-, Inter- and Transdisciplinarity*, vol. 12, no. 2, 2012, pp. 5–18.

Peters, John Durham. *Speaking into the Air: A History of the Idea of Communication*. University of Chicago Press, 1999.

Premnath, D. N. "The Concepts of Rta and Maat: A Study in Comparison." *Biblical Interpretation*, vol. 2, no. 3, 1994, pp. 325–339.

Richards, Dona. "The Nyama of the Blacksmith: The Metaphysical Significance of Metallurgy in Africa." *Journal of Black Studies*, vol. 12, no. 2, 1981, pp. 218–238.

Rigoglioso, Marguerite. *The Cult of Divine Birth in Ancient Greece*. Palgrave Macmillan, 2009.

Roberts, Don D. *Kierkegaard's Ethical and Ethico-Religious Dialectic of Communication*. Master of Arts and Philosophy, University of Illinois at Urbana-Champaign, 1956.

Serageldin, Ismail. "Ancient Alexandria and the Dawn of Medical Science." *Global Cardiology Science & Practice*, vol. 2013, no. 4, 2013, p. 395.

Skeat, T. C., and E. G. Turner. "An Oracle of Hermes Trismegistos at Saqqara." *The Journal of Egyptian* Archaeology, vol. 54, 1968, p. 199.

Vergados, Athanassios. *The "Homeric Hymn to Hermes": Introduction, Text and Commentary.* vol. Bd. 41. De Gruyter, 2013.

Wiredu, Kwasi. *Cultural Universals and Particulars: An African Perspective.* Indiana University Press, 1996.

Wright, Sarah. "Ethical Seductions: A Comparative Reading of Unamuno's El Hermano Juan and Kierkegaard's Either/Or." *Anales De La Literatura Española Contemporánea*, vol. 29, no. 2, 2004, p. 119.

Yoyotte, Jean. "Pharaonic Egypt: Society, Economy and Culture." *General History of Africa Vol II: Ancient Civilizations of Africa*, edited by G. Mokhtar. UNESCO, pp. 112–135.

4 The Universal Moral Ideal of Maat

Figure 4.1 Papyrus of Entywny (1. 521.5cm., max. ht. 35.5cm.).

In contrast to other classical approaches to ethics, Maat was more than a moral ideal in Kemet; it was a philosophical, spiritual, and aesthetic ideal. Today, it would be difficult to envision a set of principles that can be so widespread and so entrenched in a culture that it could provide the axioms to what is good and what is beautiful simultaneously. Yet, such is the case with Maat. In this chapter, I endeavor to present Maat as a universal, all-encompassing moral ideal that transcended the real morality to a broader cultural paradigm. In doing so, I will be drawing parallels to Kant's deontological approach to ethics in an effort to continue to make Maat conversant with other major approaches to ethics and communication ethics. However, I wish to make clear that Maat does stand on its own as a complete universal approach to ethics that must be studied in its own terms. Thus, the comparisons that will be made to Kant are solely for the purposes of further differentiating Maat from other canonical approaches. I have chosen Kant because his approach to ethics has both parallels and significant differences to Maat, especially in the realm of aesthetic judgments that can help us make sense of the confluence of morality and aesthetics.

Although Kemites maintained a cultural local focus of interest, following the ethical imperative of "Know Thyself", there is no question, Maat as an approach to ethics is universal. How is this possible? Think about the rules of chess. The rules of chess apply universally to all individuals who play the game and do not apply to those who do not. In the now decades-long debate about whether universal ethics are still applicable to ever more diverse and multi-cultural societies, there hasn't been enough clarity on what the term universal means as applied to ethical principles.

Today, ethicists like Seyla Benhabib have contributed a great deal to a more thorough understanding of communication ethics. In examining, whether universal claims can be made in a post-enlightenment world, she cautions us to take a more critical look at "otherness", particularly at the ethical problems that arise when we theorize about a "generalized" versus a "concrete" other (51). She proposes an interactive universalism that is dialogic, as a way to achieve better knowledge of the self that, in turn, will enhance our ability to empathize and relate to others ethically. The "concrete" other for Benhabib is a key factor for obtaining the necessary epistemic information to assess whether our moral situation is "like" or "unlike" that of others. One of her main concerns is the idea that ethical thinking cannot be done in isolation, in a vacuum, or worse, limited to an autonomous self, "a narcissist who sees the world in his own image, who has no awareness of the limits of his own desires and passions and who cannot see himself through the eyes of another" (156). For Benhabib, communication, discourse, and meaningful interaction/dialogue with others are the centerpiece of the ethical paradigm.

She contrasts an interactive moral agent to the transcendental individual, the autonomous rational agent she sees in moral theories espoused by enlightenment thinkers, particularly Immanuel Kant. In her view, "Kant's error was to assume that I, as a pure rational agent reasoning for myself could reach a conclusion that would be acceptable for all times and all places" (163). In her reading of Kant, she compares the autonomous rational agent to a geometrician, who, separated physically from other geometricians, is still able to arrive at the same conclusions as others solely based on her capacity to reason.

In a similar fashion, Daryl Koehn juxtaposes the (ostensibly silent) Kantian universality test with a discursively based communicative ethics. Koehn, in particular, interprets the role of the categorical imperative in Kant as a quasi-mathematical form of ethical reasoning. She states that "Ethical reasoning is not distinguished by any effort at consultation with others but rather by a desire to state, defend, and apply ethical principles" (2). Similarly, Benhabib charges Kant with relying on legalistic universalism and argues that: "The first step in the formulation of a post-metaphysical universalist position is to shift from a substantialistic to a discursive, communicative concept of rationality" (5).

Both of these ethicists see their projects as confronting a rationalistic account of ethical communication that has largely discredited the universalist project due to its perceived disregard for the sort of moral reasoning that is done taking into account context (or material realities), thus excluding issues of gender, race, and communitarianism from its purview of concerns. This is an important critique that is echoed by other contemporary ethicists and there can be little doubt that as Clifford Christians argues in his "Sacredness of Life" essay, "the manner in which, race, age, gender, class, disabilities, and ethnicity is represented in language provides the possibility for a just socioeconomic order" (4). I agree with Christians on this point. However, I believe that the Maatian approach to ethics is one example of the way in which both the concepts of reason and abstraction need to be revisited, so that (1) we can thoroughly explore the full implications Kant meant to describe with this construct; and (2) we reimagine these two concepts in Maat ethics to help illustrate other aspects, namely internal aspects, of ethical communication that may have previously been neglected in our analysis.

Re-Enlightening the Enlightenment Transcendental Individual

I find the present interpretation of Kantian autonomy and reason to be somewhat narrow in scope and limited in its interpretation. For Kant, autonomy is the ability to live by one's own law. But is the autonomous individual by necessity also selfish and unconcerned with others? And what would "one's law" mean when we are talking about ethical and moral principles? Is it possible that behind this apparent disproportionate concern with self-determination, this concept of autonomy, Kant might actually be alluding to his greater philosophical project of exploring the mechanisms by which individuals can harmonize with the world of possible experience? If this description of the Kantian ethical project sounds suspiciously like the way in which I have presented Maat, I hope that by the end of this chapter, I've succeeded in illustrating the ways in which deontological ethics do share similarities with the Maat approach but also to have shown the many ways in which they diverge from one another.

I'll start with three obvious questions: Can an autonomous individual, in the Kantian sense, embrace communally (externally) agreed upon norms? And, if so, what type of communication practices might facilitate individuals freely and authentically contributing to the general consensus of society on morality? Can this contribution be considered an extension of their autonomy? The answer is yes to all three questions for Kant and in a Maatian context. One's own law can indeed, and in Kant's case, is likely a match for existing communal norms behavior. But this kind of alignment requires both a meaningful interaction with other ethical agents and the sort of reflective practice that engages a Kantian form of reasoning through the categorical imperative. According to Clifford Christians, "The first issue (in ethical theory) is to retheorize theory away from epistemic certainty" (Ethical Theory 9). I think the concept of reason, as described by Kant, fits this mandate although it may seem the unlikely at first.

Regrettably, reason is often viewed as a metaphysical, arbitrary set of laws. Yet, as we have seen in a Maatian context, reason can and had been directed toward the principal end goal of sustaining and improving the conditions necessary to survive on this planet. This focus on survival has not always been evident throughout history and across cultures, but in the case of Kemet, it is not only possible but has been recorded for posterity. Similarly, Kant views reason as an organizing principle by which we can make sense of possible experience. In Kant's case, he means without relying on absolute certainty, as it is foreclosed to us in the noumenal world. However, a lack of absolute certainty about the nature of things-in-themselves, which includes the nature of human beings, doesn't translate into a call for amoral or immoral human behavior.

For Kant, reason is a protonorm. A protonorm, as such, is presupposed. It is an agreed upon frame of reference. In fact, one of the duties of reason for Kant is to always try to synthetically (through reason) transcend the world of experience to find purposiveness in our relationships to one another (Formula of Humanity). In Kant's logic, without reason, there cannot be purpose, and without purpose, we are left with an infinite number of empirical facts and interpretations of the world of experience that may tell us something about the physical reality of our existence, but tell us little about the way we ought to behave in order for us to thrive individually or collectively. For Kant, there are many physical laws in the world of possible experience that can, indeed, be quantified and known through the use of mathematical and scientific principles.

Here again, we see a parallel between Maatian and Kierkegaardian ethics. There is agreement in these three views that human knowledge is limited to the world of experience (keep in mind that the spiritual world is a very real and objective part of the human experience in Kemetic ontology and other African and world cultures). Furthermore, there is agreement that both the physical world and the spiritual world (one's own will, consciousness, ba, maueutic artistry, righteous speech) provide opportunities for exploration, experimentation, and discovery of one's own abilities and dispositions.

In other words, there are just as many phenomena for which scientific and mathematical knowledge is simply inadequate if we are trying to understand or evaluate its functioning and our relationship to them as human beings. Like Kwasi Wiredu, Kant thinks that morality is universal because it is necessary for the existence of a human community (202). Furthermore, the justification for why morality (ethics) has to do with reason is because "whilst only understanding is necessary to comprehend the mechanical linking of states of matter . . . in order to perceive purposes one needs reason" (Bowie 46). Take, for example, Kant's description of reflective judgments. These judgments allow us to make ethical decisions in the absence of a given law.

Reflective judgments are illustrative of Kant's insistence on the realm of ethics being one of duty more than a matter law. It bears repeating, given the widespread misconceptions, that it is a mistake to imply that for Kant everything about ourselves and the world is "knowable". What we have in place of absolute certainty is reason. Reason affords us the ability to: (1) Make observations; (2) identify patterns through those observations; (3) make inferences about those observations; and (4) make judgments based on those inferences. More broadly, this is describing the process of abstraction necessary in order to practice moral reasoning. Abstraction, it turns out, places a vital role in our ability to behave ethically.

But one might ask, what about Benhabib's concrete other? What sort of relationships could possibly be established through reason since it relies so heavily on abstraction? "Kant is careful to distinguish between abstraction and analysis: abstraction separates a single quality from a composite whole, whereas analysis distinguishes between all present qualities." (Caygill 41). In other words, abstraction is a mechanism by which we are able to conceptually isolate one component out of a whole for the purposes of comparison or identifying a pattern. This is a particularly useful tool in ethics. After all, the Kierkegaardian model of ethical communication described in the previous chapter calls for a space as well as a type of reflective mode of communication where individuals are able to make their own thinking an object of inquiry.

Thus, the process of abstraction is vital to a universal approach to ethics, like Maat, for three main reasons: (1) As stated previously, the tool of reason, which is based on our ability to abstract from a whole (a universal principle), is necessary in order to provide the purposiveness, the protonorm that can enable and foment ethical discourse; (2) although colloquially, and even some contemporary ethical theorists understand the idea of abstraction as a way to relate to the world (and others) in a disembodied, disconnected, and inflexible way, philosophical abstraction is based on the idea that the best inferences made from abstraction are based on the best possible knowledge of the particular, nuanced, historically situated components of the whole. Maat is paradigmatic of this practice as it placed such emphasis on the study of natural processes and laws of nature, on the individual acquisition of knowledge for the purposes of gaining a better understanding of Maat's ultimate ends. As one scholar puts it:

> In a word, doing Maat among humans is the fundamental way of doing Maat for God. This does not diminish the importance of specific duties to nature or specific duties to God. But it does give greater stress to moral practice in the human community which is never in isolation or without direct or indirect implications for relations with God and nature (Maat 311).

Communication as Key to Survival

As the originators of philosophical thought as well as its formal study, Kemites understood the usefulness and proper place of abstraction in both philosophy and ethico-religious thought. That the study of philosophy began in the continent of Africa, today we have little doubt. Scholars have been noting and debating the new archeological evidence and as a result, a corrected version of our human intellectual history has been shaking the academy in almost every field of study. A new paradigm of intellectual justice has been slowly replacing the old "Aryan" paradigm as scholar Martin Bernal termed it in his now famous Black Athena: *The Afroasiatic Roots of Classical Civilization: The Fabrication of Ancient Greece 1785–1985. Vol. 1:*

> After the 5th century BC – the only period from which we have any substantial knowledge of them – the Ancient Greeks, though proud of themselves and their recent accomplishments, did not see their political institutions, science, philosophy or religion as original. Instead they derived them – through the early colonization and later study by Greeks abroad – from the East in general and Egypt in particular (261).

The fact that Kemites had been practicing abstraction with regard to ethics is evident in their use of neteru as illustrative in the key forces of nature that human beings must study, appreciate, and with which human beings must engage. The best forms of abstraction are based on exhaustive and thoughtful consideration of the nuances contained in the whole. It is one thing to look superficially at a whole and extract one or a set of generalizations that misrepresent the whole. It is another to collect through generations observations and knowledge of a particular place and population and then deduce from those inferences a set of values directed at accomplishing collective end goals. Thus, contemporary ethical theory should not dismiss the practice of wholesale abstraction because at times it may be performed poorly. I maintain that the process of abstraction is valuable to the discourse of ethics because it provides a unique exercise where groups and individuals can test the continued validity and relevancy of universal ethical principles.

Take, for example, the ecological components of Maat. The process of abstraction allows individuals to negotiate their own role in relation to their family, their community, and nature at large. In order to understand the logic underpinning "as above so below", it is crucial that individuals exercise their reasoning ability to understand the inner-workings of above, to understand what is meant by below, and be able to figure out the directives this yields for them individually and as part of a collective. In this excerpt from the *Spell from Passing Akhet to the Sky* (Part of the Pyramid Texts of Pepi I, third King of the sixth Dynasty (ca. 22892255 BC)), we see an expression in poetic form of the way in which Kemites understood their relationship to life-giving nature even as they transition into the future life (afterlife) and we get a sense for the complete immersion they felt in earth's elements. This is clearly not a relationship of worship but of kinship:

> I receive the offering slab and manage the "gods-mouth" altar. I shoulder the sky with life and support the earth with happiness:

this right arm of mine shoulders the sky with a staff, this left arm of mine supports the earth with happiness. I always find a fare for myself because the abomination of Summoner, the doorkeeper of Osiris, is ferrying without a toll having been paid to him. I receive for myself my air of life. I inhale happiness and become sated with god's offerings: when I have breathed the air of my abundance, the north wind, I become sated among the gods (Allen 159)

Again, the difference between a Maatian view of ethics and a deontological one is that Kant discards the idea that there is something driving our knowledge of the world, that is, in itself, outside of a certain kind of discourse of experience, for as human beings we lack the "God's eye" point of view to verify some of our most basic claims to knowledge outside of our own experience. However, a Maat views human beings as partaking in Ka, as possessing aspects of the divine creative force, and as a result it is further incumbent upon individuals to pursue complete self-actualization through the process of becoming "true of voice" (*Maa-Kher*).

Still, self-actualization can only occur when we have the time and resources to do and speak Maat with some regularity. Furthermore, if one understands the self as interdependent with a community that includes nature, appreciating and highlighting the value of other species and landscapes in word and deed become part of one's ethical responsibility. Thus, the ecological knowledge built into Maat cannot be overstated. The view itself is the product of millennia of distilled ecological and intellectual wisdom that sets Maat apart from other classical ethical views. In the words of Senegalese Anthropologist and Historian Cheikh Anta Diop:

The history of humanity will remain confused as long as we fail to distinguish between the two early cradles in which Nature fashioned the instincts, temperament, habits, and ethical concepts of the two subdivisions before they met each other after a long separation dating back to prehistoric times. The first of those cradles, ..., is the valley of the Nile, from the Great Lakes to the Delta, across the so-called "Anglo-Egyptian" Sudan. The abundance of vital resources, its sedentary, agricultural character, the specific conditions of the valley, will engender in man, that is, in the Negro, a gentle, idealistic, peaceful nature, endowed with a spirit of justice and gaiety. All these virtues were more or less indispensable for daily coexistence (97).

Interestingly, the question of whether the reliance on reason, syllogistic arguments, or abstract principles precludes the possibility of a meaningful, concrete account of material reality has been debated and even found some unlikely allies in some of the most important advocates of an ethics situated in socio-historical reality. One prominent example is Karl Marx, who embraced the concept of abstraction in the service of his theory of materialism.

> Following Kant and the scholastic tradition, Marx restricts the use of abstractions to revealing something specific through comparison of "common specific qualities", in his case those between different social formations and the specific qualities they appear to have in common (Caygill 42).

Therefore, the concepts of abstraction and ethical theories that account for material reality are not necessarily mutually exclusive. As ethicists, we must think of abstraction as a communication tool, and as a mechanism that, while certainly susceptible to misuse or poor execution, can and should be helpful in identifying commonalities that can serve to bring together a wide range of individuals and interests for the sake of their collective flourishing. So far, I have juxtaposed two universal approaches to ethics that culturally and chronologically bare no commonalities. These two views could not be further apart in their formation or current cultural capital. In pointing out key similarities and differences, I have highlighted the ecological and spiritual aspects of Maat as they are absent from the enlightenment view of deontology popularized by Kant. I will now turn to the question of whether universal approaches to ethics, and specifically those like Maat that incorporate aesthetics as a dimension of morality, still have a place in our modern sensibility.

On Reasoned Pleasures

Although for Kant the principle underlying judgments in general are perplexing, aesthetic judgments, in particular, illustrate the complexity and importance of the function of judgments in universal ethics. In the *Critique of Judgment*, Kant argues that there are a priori principles that undergird aesthetic judgments, as much as there are such a priori judgments on the basis of reason. However, while Kant stresses the importance of investigating critically the principles of judgment, or more appropriately aesthetic judgment, he also contends that they do not contribute to the "knowledge" of things. In this reading of the role of aesthetic pleasure in morality and wisdom, Kant diverges from Maatian ethics.

Part of the reason for this is because for Kant although aesthetic judgments share a universal basis, the feelings that are bounded with aesthetic judgments are not and should not, by definition, be knowable in the way in which, say scientific knowledge is possible. Aesthetic judgments, inherently, are linked to finding/discovering purposiveness in our world, and therefore the communication involved with these judgments should function as a medium as opposed to a science. Hence, because judgments belong to the cognitive faculty, and because they uniquely reference our feelings of pleasure and pain, he maintains that no *Critique* would be complete without an elucidation of this very distinct and puzzling phenomenon. Hence, he declares the *Critique of Judgment* the end of his whole critical undertaking (6).

Aesthetic judgments as judgments proper are made through our natural faculty of taking the particular under the universal (abstraction). For Kant, there is "something in our judgments upon nature which makes us attentive to its purposiveness for our understanding" (24). Put another way, there seems to be something innate in us that seeks to organize and harmonize laws and phenomena that might seem dissimilar at first. More than a human predilection, this purposiveness we search for, this ordered whole view in Maatian terms, is the only means that we have available to know nature.

This, however, is not to say that as humans we have no choice but to settle for a second-rate sort of truth; rather, I think his point in the first critique was to say that a knowledge of nature, not as an object in itself, but as a perceived phenomenon, is good enough since as humans we too are part of nature. Here, we see a bit of resistance on Kant's part to Cartesian dualism and this is not the place where he, at least, attempts to refute Descartes. In Kant's concept of human beings, he describes us as heteronomous and concedes that man is essentially half-human and half-animal and for this reason, it is vital that the part of ourselves capable of reason is cultivated and promoted.

Kant is careful to distinguish the idea of purposiveness from purpose proper. This is a key point since he wants to maintain the point he makes in the first critique about our inability to know objects outside of our cognitive structure. For Kant, there is the noumenal and the phenomenal world, and the latter is the one that we are able to cognize and understand. So there is always a little qualification present when we speak of purpose in nature in Kantian thought. Some people view this position as a precursor to more constructivist positions, or as Hilary Putnam views it, an internalist point of view. But the idea remains that although to know the purpose as such of an object may be beyond the scope of our knowledge, or more importantly, beyond the scope of our perception, in purposiveness of form, we find how the particular object fits within our conceptual framework.

Here, we have another contrast to Maatian ethics in so far as cosmic substances are considered sacred in the Kemetic worldview. In a Maatian ethical framework, the distinction between purpose and purposiveness is bypassed for a generalized understanding of nature, which includes human beings, as always already purposeful as a manifestation of the divine. No further explanation or category is needed to justify our purpose in the cosmos. Nature, and humans as part of it, exists for the purpose of promoting life in all its forms. Evidence of this can be found in the way Kemites conceptualized the role of "king". In Medu Netcher, the word for "pharaoh" is "great house" (Butrick 68) and the king was expected to embody the principles of Maat as well as the life force that was so important to the people of Kemet:

> The monarch, the revered being par excellence, was also supposed to be the man with the greatest life force or energy. When the level of his life force fell below a certain minimum, it could only be a risk to his people if he continued to rule. This vitalistic conception is the foundation of all traditional African kingdoms, I mean, of all kingdoms not usurped (Diop 124)

At the beginning of the chapter, I mentioned that Maat functioned as both a moral ideal and aesthetic ideal in Kemet. So what is meant by ideal and how does it relate to morality? Are Maatian ethics preoccupied with matters of taste, of beauty, of pleasure? What, if any, role do feelings of pleasure have in ethics? Is there such a thing as a reasoned pleasure? The short answer is yes. For readers familiar with the cannon of ethics, John Stuart Mill begins with treatise on utilitarianism by differentiating higher pleasures from lower pleasures and he believes that higher pleasures are those produced by intellectual and moral activities. A similar approach is present in Maatian ethics as aesthetic pleasure (a higher pleasure in Mill's view) is derived from the harmonious and balanced design of spaces and structures for example (Cannon-Brown 13).

However, it comes to Kantian aesthetic judgments; these are mostly concerned with judgments of taste, of judging the beautiful. For Kant, our pleasure is bound up with our apprehension of forms of objects of intuition (Critique of Judgement 26). These forms of objects of intuition, however, are immediately combined with our feelings of pleasure since we can't help but use representations of the object for cognition and this representation, this apprehension, is always met with innate pleasure. Hence, this pleasure that we feel is not, according to Kant, associated with knowledge or acquaintanceship with the object itself (since we are not going to classify it or identify it in any way that is relevant to knowledge), but is more of a subjective pleasure that is not linked to a concept but to a feeling, a satisfaction in our ability to perceive in possessing and using our faculties. Kant is here careful to clarify that although judgments of taste are not judgments of cognition (that would require the use of concepts), these judgments are still an important part of the cognitive process. Put another way, the pleasure we feel in the representation of a perceived object does not reference the object as much as it references the subject who is caught up in the pleasure of representing it. This is a faculty and a pleasure that is common to everyone and hence, everyone has the faculty of judging the beautiful.

One important caveat, however, is that although everyone is capable of judging the beautiful, which, for Kant, requires the implied agreement among all human beings, beauty as such can only be experienced subjectively. In other words, there is no set of rules that defines what is to fit within the category of beautiful, for if there was and a concept was created to determine so, this would cease to be an aesthetic judgment to become a cognitive or logical one. So, this leaves us to ask: How can one ever be sure that what one considers beautiful really is? In the absence of a concept, can we ever communicate to others these judgments of beauty? Kant's answer to these questions is yes. Beauty is universally communicable and yes, there are ways to appreciate and recognize beauty in a way that is not corrupted by more passing inclinations such as pleasure or matters of personal taste, all while remaining a deeply subjective call. Hence, it seems to me that one of most fascinating aspects of Kant's exposition of aesthetic judgments is this idea that is subjective and can be shared, or universally communicated.

In short, Kant implies that aesthetic judgments operate according to rule (since everybody can recognize it and communicate it) but not according to concepts. The reason for this is that aesthetic judgments are more about how the subject harmonizes with the universal, through this pleasure in representation, than about any sensation of the object or "reference to any concept which anywhere involves design" (Critique of Judgement 27). In other words, this harmony that is achieved has less to do with how the object harmonizes with concepts than highlighting the purposiveness of the cognitive faculties of the subject (Critique of Judgement 27).

In this section, I've juxtaposed the aesthetic dimensions of Maat to Kant's for the purposes of contrasting the two views and continue to put Maat in conversation with the current cannon of ethics. Kant insists that aesthetic judgments are communicable universally by means of some implied universal agreement. Kemetic thought on ethics parallels this view in so far as the object of beauty, morality, and communication is to access a universal consciousness that can be most efficiently accessed through the teaching of Maat. Maatian ethics decenter the individual as the nexus of meaning and value and locate it instead on a consciousness that though real, present, and constant is up to the individual to embody in thought, word, and deed.

Maatian Symmetry and Balance as Visual Ethics

Any person vaguely acquainted with the wonders and achievements of Kemetic culture probably pictures countless beautiful structures, pyramids, temples, sculptures, and paintings that defy the imagination in terms of their grandeur, resilience, and scale. Someone more familiar with Kemetic culture would also know that Kemites refined geometry and architecture into an art form. Much work remains to be done in the area of the aesthetic dimensions of this moral ideal but one scholar has noted that while the word "nefer" (beauty, goodness, the good) was one of the most used words in the Medu Neter language, they did not leave a definition of what the beautiful is (Cannon-Brown 2). This seems peculiar given the meticulous treatment of so many other philosophical matters, including communication. Was this done on purpose? Is aesthetic judgment yet another area of confluence between Kant and Kemetic ethics? What we do know is that the term nefer was also used to describe speech. "Medu nefer" (mdw nfr) meant good speech, beautiful speech, which alerts us that Maat had a criteria by which to judge, at least some modalities and qualities of speech. In the context of rhetoric, medu nefer referred to the virtues of exercising self-restraint, logic, humility, and respect when speaking, all of which meant to reflect the overarching ethical values and aesthetic of Maat (Karenga 217). This gives us a good idea if what Kemites considered beautiful and good content when it came to speech, but what of the form?

Form was of great importance as well. The word *isfet* meant "evil, imbalanced" and the reader might be surprised to realize that some of this terminology is still operative today. Think about the phrase of being "ugly on the inside". This contemporary phrase encapsulates the Maatian preoccupation with balance and harmony in the way a person's inner thoughts may be "imbalanced" or how their behavior may create "disharmony" or conflict with other. Here is one more example of how morality and aesthetics not just did but still are used together in judgments and how this admixture manifests itself in language.

A final point on the issue of the universal ability to recognize beauty is that according to Kant, similar to logical judgments, aesthetic judgments can be divided into empirical and pure. We can only recognize the form of the representation as beautiful. Empirical judgments of beauty, like those regarding such things as the beauty of a particular color or tone, can assert the beauty of an object or of its representation, but these are also more prone to be informed by socialization practices as opposed to being instinctive. Pure aesthetical judgments, however, only assert pleasantness or unpleasantness based on their form (Critique of Judgement 59). These judgments are more directly linked to feelings of pleasure, since they don't depend on more ephemeral traits such as charm to get in the way of judging something that is purely beautiful. This is a rather daring position, since by implication Kant is suggesting that even something which could be considered ugly can really be beautiful.

In contrast, Kemites did not seem to share the same preoccupation with purity of ethical judgments. Here, it is important to note that as an ethics, Maat places a strong emphasis on listening practices. These listening practices are conceived as necessary to the proper harmonization of the individual to her environment, human, and natural all the same. Thus, it must be noted that while the morality and values that drive Maatian communication differ from Kant's, the Maatian framework does, in fact, treat its own categories and criteria with the same meticulousness and care. Evidence of this is the vocabulary that Kemites developed to study and differentiate different aspects of Maat as it relates to ethical speech practices. For example:

> The concept rekhet (written with the hieroglyph for abstract notions) means "knowledge," "science," in the sense of "philosophy," that is, inquiry into the nature of things (khet) based on accurate knowledge (*rekhet*) and good (*nefer*) judgment (upi). The word upi means "to judge," "to discern," that is, "to dissect." The cognate word upet means "specification," "judgment," and upset means "specify," that is, give the details of something (Obenga 33).

Though differing in emphasis and ontology, both Maatian ethics and Kantian deontological ethics illustrate the type of ethical theory that while universal in scope, take some precautions to ground morality in a particular place and time. While Kant relies on the categorical imperative as the universal test for ethical decision making, Maat ethics provides a set of principles and mechanisms no less sophisticated and nuanced than Kant's. But is intricacy the right kind of standard by which to judge the relevance and current usefulness of an ethical theory?

Conclusion: Unaltered by the Winds of Change

While interactive universalism and other dialogic ethical models are based on principles of trust, care, empathy, truthfulness, among others, it seems that there are (or at least could be) other vital components of an ethical life, such as beauty, that can add dimension and depth to what is considered right or good. Contemporary ethical theorists like Martha Nussbaum privilege the notion of well-being over personal or internal satisfaction (as it is reflected in her theory of adaptive preferences), while other ethical theorists like Kwasi Wiredu have expressed their worries about the role of non-conceptual truths in ethical discourse, especially as it relates to intercultural understandings of truths.

Maatian ethics provide directives for connecting the individual to a universal consciousness in the sense that its wisdom transcends species and time. In the words of Hegel, we become self-conscious when we reflect back upon ourselves and come to understand the "I" that we previously thought as only pertinent to ourselves as a universal "I" (122). This is when we begin to see ourselves reflected in the objects and subjects that make up the world around us. This means that we become engaged in a more practical type of consciousness that has returned from its "otherness" unto itself. That we have turned the mirror back on ourselves and have begun to discover that there is a lot we share with others (human and otherwise) and that rather than different and separate from others we are the embodiments of something larger than us. In the Hegelian as well as the Maatian worldview, the subject/object dichotomy becomes blurred and trivial, irrelevant. Thus, the activity of self-consciousness consists in Hegelian terms as negating the first negation of consciousness. But in more Kierkegaardian terms, "that the task is not to move from the individual to the universal but from the individual through the universal to reach the individual." (Kierkegarrd 90).

Now, since for Hegel our consciousness as a whole is part of the history of the world and for Kemites our consciousness encompasses the wholly sacred and divine, we must be able to understand the contents of our experience in this world as such. Therefore, we must first learn not to want to privilege subject over substance and then we must act accordingly. Everything is connected for Hege and in Maatl. Hence, consciousness must involve forms of knowing (of understanding our experience in this world) that actually make a difference on how things go on. Thus, unless thought matches up with our actions, *Geist* can never become actualized (Hegel 240). The difference, of course, between Hegel's view and a Maatian view is that the Ka is always already actualized and it is individual people and groups who risk missing out full actualization and not the *Ka* itself. Maat is here to guide our steps toward this self-actualization and not to guide the Ka.

Having gone over the development of Maat in pastoral and agricultural settings, having looked at the cosmology and theories of the soul that helped shape and manifest themselves in Maat, and having touched on the scope of Maat as a metaethics of both human ethics and aesthetics, we are now ready to explore the many ways in which Maat offers a specific, fully realized, and internally cohesive theory of communication. In many ways, a Maatian approach to Maat can be summed up as "communicating to posterity". The next chapter takes a closer look at the specific communicative aspects of Maat, including many examples from literature, funerary texts, and poetry to evidence said cultural and linguistic expressions of Maat.

Bibliography

Allen, James P. *The Ancient Egyptian Pyramid Texts.* no. 38. SBL Press, 2015.

Benhabib, Seyla. *Situating the Self: Gender, Community, and Postmodernism in Contemporary Ethics.* Routledge, 1992.

Bernal, Marin. *Black Athena: The Afroasiatic Roots of Classical Civilization: The Fabrication of Ancient Greece 1785–1985.* vol. 1. Rutgers, 1987.

Bowie, Andrew. *Aesthetics and Subjectivity: From Kant to Nietzsche.* Manchester University Press, 1990.

Butrick, Leifa. *Hatchepsut: The Female Pharaoh.* vol. 6, no. 2. University of Wisconsin-Milwaukee, 1997.

Cannon-Brown, Willie. *Nefer: The Aesthetic Ideal in Classical Egypt.* Routledge, 2006.

Caygill, Howard. *A Kant Dictionary.* Blackwell Publishing, 1995.

Christians, Clifford "Ethical Theory in Communications Research." *Journalism Studies*, vol. 6, no. 1, 2005, pp. 3–14.

———. "The Sacredness of Life." *Media Development*, vol. 2b, 1998, pp. 3–7.

Diop, Cheikh A., Mercer Cook, and ProQuest (Firm). *The African Origin of Civilization: Myth or Reality.* Lawrence Hill Books, 1974.

Hegel, G. F. W. *Phenomenology of the Spirit.* Translated by A. V. Miller. Oxford University Press, 1977.

Kant, Immanuel. *Critique of Judgment.* Translated by J. H. Bernard. Hafner Publishing, 1951.

———. *Critique of Pure Reason.* Translated by Norman Kemp Smith. McMillan, 1965.

Karenga, Maulana. *Maat, the Moral Ideal in Ancient Egypt (African Studies).* Taylor and Francis, 2004.

———. "Nomo, Kawaida, and Communicative Practice." *The Global Intercultural Communication Reader,* edited by Molefo K. Asante, Mike Yoshitaka, and Yin Jing. Routledge, 2008, pp. 211–225.

Koehn, Daryl. *Rethinking Feminist Ethics: Care, Trust and Empathy.* Routledge, 1998.

5 Communicative Dimensions of Maat

Speech and Silence

As a distinctive and complete classical approach to ethics, Maat has four distinguishing features from other classical approaches: (1) It conceptualizes speech as a radically creative act; (2) it promotes the use of logical and scientific language; (3) it places a unique emphasis on listening as the enabling communication skill; and (4) it conceptualizes silence as a powerful form of communication. The discussion that follows will proceed in this order while incorporating texts from the Kemite people so as to allow them to tell much of the story as possible.

The illustrations included here come from a wide range of sources. While the reader may be most familiar with funerary texts, that is script that was found inside or in the vicinity of an ancient burial, archeology has now given us access to a wider range of writings that include love songs, poetry, stelae, legal documents, short stories, biographies, anecdotes, proverbs, and the inclusion of such literature will necessarily push back against the depiction of ancient Kemites as death-worshiping people. We now know that the opposite is true. Kemites were life-loving, life-focused people and the fact that some of the first documents archeologists were able to retrieve from the earth came from tombs buried under the sands is more a matter of circumstance than a complete picture of the values, beliefs, and customs of these ancient people.

Speech as a Radically Creative Action

There are four factors associated with why the ancient Kemites associated speech creative action. The first is found in the cosmological account of *Ptah* having thoughts of the world in its heart and then issuing the first authoritative utterance when the universe was created. In this cosmology, speech communicates what is in one's heart and when that occurs, amazing things happen. The second reason has to do with the way in which the Medu Netcher symbol for mouth is also interpreted as opening one's consciousness. The third reason can be found in the pages of the *Book of the Coming Forth by Day* (often mis-titled the Books of the Dead). Among its pages, there is an account of the ceremony of the opening of the mouth, which has great spiritual weight and was well-known among Kemites. The fourth and final reason why Maatian communication ethics can be said to be distinctive in its view of speech as a radically creative act is the association of speech and magic. All of these factors contribute to an approach to communication ethics that can be readily recognized as unique and offering something new to the cannon that can help shed further light into the intersection of morality and communication.

Kemetic cosmology begins with the word. Ptah is said to have ordered the world into existence through the use of divine speech (Assmann 19). It is worth noting that the origination of the impulse to create the universe came from Ptah's heart and it was only after the heart generated these ideas and the mouth communicated those ideas, making them into a reality. An excerpt from a hymn to Amun-Re dating from around 1400 BCE tells about the supremacy of speech as a major creative act in song:

> He came forth as self-generated,
> **all his limbs speaking to him**
> He formed himself before heaven and earth came into being the earth being
> in the primeval waters in the midst of the "weary flood"
>
> You have started to create this land
> **to establish what has come from your mouth (= the gods)**
> You have raised heaven and kept earth down
> to make this land wide enough for your image

I ask the reader to please note that the author writing these pages does not agree with the past Egyptological practice of translating the word "ntr" as "God". More current scholarship uses the terms "force of nature", "divinity", "deity" (West 237) as a replacement in order to be consistent with the fact that Kemet exhibits a robust spirituality as opposed to a formal religion. As one scholar explains:

> In order to challenge European religious mythology in the twenty-first century, Afrikan people must relocate, redirect, and re-orient themselves to their original status as a spiritual people in the B.C.

era in the Nile Valley, Kemet (Egypt). They must also recognize the poignant difference between religion and spirituality: Religion represents the deification of a people's cultural experiences, politics, and political power control intent. Spirituality represents a direct connectedness/inter-relatedness with nature, the cosmos, the universe and that spiritual God-force, Amun-Ra, "the giver of life." In ancient Kemet, man was the reflective image of the cosmos, universe, and Amun-Ra (Natambu 366).

Thus, rather than an impediment to the study of Kemet, such issues of translation are brought to the fore to further illustrate the conflicted history of the field of Egyptology (Karenga 203) and the success modern scholarship has had decolonizing it. "A major tendency in Egyptology is to look for and discover the origins of evil in human nature in search of parallels with Jewish and Christian teachings" (Karenga 203). As a communication ethicist, I am sensitive to the plight of colonized people worldwide and see it as my ethical duty to listen to their voices without the filter of imperial power. Thus, the remainder of this essay will exhibit the same commitment.

Mirroring the very first act of creation, Maat promotes speech as a way to create harmony, balance, reciprocity, truth, and life through the use of speech. Here, we have a communication ethics that considers speech as sacred, divine. A divine force that manifests itself through a divine means, human beings. The meaning of the language of Medu Netcher is "divine speech" (Alkebulan 23) and Kemites conceptualized language, as the name implies, as a form of divine power that should not be taken lightly. Every time human beings open their mouth, in a sense, they are recreating the very first act of creation and it is this fact that gives communication practices an air of ritual, of the remembrance of that first divine utterance.

Second, speech is seen as a creative force in the Maatian approach to ethics through the use of the symbol of the mouth:

Figure 5.1 Medu Netcher symbol for mouth.

Though its symbol only comprised the simple outline of lips, the symbol for mouth also represents consciousness in the Medu Netcher language. Having reviewed the cosmological account of how Ptah created the world, it should come as no surprise that the symbol for mouth would also be associated with the ability to create a new reality, with the ability to infuse others with divine energy and self-awareness, and with the ability words have to set in motion great change. An open mouth, in short, is able to bring about truly magnificent events, when used wisely.

Third, Kemetic culture did share a cultural belief of life after (mortal) death and as such took some care to equip their loved ones with symbols of the kind of life they wished for them upon their departure. One scholar describes the ceremony of the opening of the mouth as "arguably the most commonly attested Ancient Egyptian ritual" (Ayal 113). This ceremony was part of an ancient mummification procedure in which priests opened the mouth, nose, and eyes of the deceased so that the body could breathe in addition to receiving the sustenance (food, drink) a body normally needs, while the "ba" (individual soul, personality) made its way back from the *Duat* (often called underworld) after completing their journey of final judgment. Furthermore, one scholar describes an additional important reason why Kemites thought it necessary to open their loved ones' mouth after death: "Then follows an important ceremony, the Opening of the Mouth, to make sure he is restored to the power of speech: 'May you give me my mouth with which I may speak, and may my heart guide me at its hour of destroying the night'" (Dimock 33).

At the end of the quote, the author is citing a passage from the *Book of the Coming Forth by Day*, plates 5 and 6. There were other components to the ritual including some animal blood incorporated into ceremony and poured on the deceased for the purposes of staging the re-birth of the loved one into a new life and form of existence. Additionally, the implements used during this procedure were the same tools used to assist in births (Ayad 113).

The ritual had been performed since the Old Kingdom but in the New Kingdom, so-called, surrogate statues, and later busts took the place of the deceased as a location where the "ba" could return. This may seem a bit "morbid" to our modern sensibility, but we still embalm our dead, and we still make flower and other offerings to them. Another reason why this is significant is because of a parallel between this rite and the Socratic method that Socrates termed "Maieutic". In the new Kingdom (empire kingdom), when statues began to replace the bodies of the deceased, this ritual became closely associated with the art of sculpture and as such, Kemites began to think of sculptors as those who help birth a new person through their careful manipulations of the raw materials. In other words, at the time sculptors took on the qualities of a midwife.

In this sense, the opening of the mouth ceremony and the idea that through the maneuvering and the careful and artful handling of an object can work to bring a person back to life is connected to the kind of "birthing" and Socrates implies when he uses the descriptor of a midwife technique of speech. In other words, in both the ritual associated with the opening of the mouth ceremony and in the concept of the philosopher as midwife, there is an assumption that through artful intervention one can pry open life in another. I mentioned earlier that there was a ritualistic association with Maat and the ceremony of the opening of the mouth as ritual does bring together two integral components of ethics: Speech and action. In this section, we've looked at multiple ways in which Kemetic culture placed a high value on the power of speech and specifically attributed to it magnificent powers of creation, creativity, and artistry. In the next section, we will take a look at the ways in which a Maatian approach to communication ethics also features an emphasis on a logical and scientific speech.

Maat as a Scientific Speech

In Chapter 3, we touched on some aspects of Maat and Kemetic society that specifically promoted logic, ecology, and a scientific approach to knowledge creation. In this section, we will look at a different set of texts that will allow the ancestors to make this point in their own words. I feel this point needs to be belabored for the reason that although Kemet still conjures images of greatness and achievement in the popular imagination, somehow questions linger about the source of their aptitude and success. From ancient aliens to stripping them of their African heritage, observers have run the gamut of possibilities to explain why these ancient African people reached such a level of scientific sophistication and technological prowess.

As a communications Professor, who is agnostic about aliens (who knows!), I am interested in the ways the Kemite value system and communication practices may have promoted or hindered their scientific endeavors. This section, then, is an exploration of that possibility. What if a highly capable and erudite society built a value system (Maat) and a communication practice that stimulated, encouraged, and rewarded scientific achievement? Wouldn't that be a better explanation for the feats of accomplishment that survive to this very day?

An excerpt from *The Instruction of King Meri Kere*, a volume in the wisdom literature (sapient texts) of Kemet addresses the "The Value of Speking Well and of Wisdom" extolling the artistry and power of good communication. In particular, this passage asserts that "speech is mightier than any fighting". This is a claim I have found in no other classical approach to communication ethics. The author is imbuing good speech with the capacity of being more impactful than crude violence. More importantly, this passage illustrates the point that in the Maatian approach to ethics, good speech (*medu-nefer*) is entwined with morality; it makes a moral demand, and reflects the principles of Maat.

> Be a craftsman in speech, so that thou mayst prevail, for the power of (a man) is the tongue and speech is mightier thananay fighting. - - - - He that is clever, him that learned attack not, if he is learned, and no (harm) happeneth, where he is Truth cometh to him fully kneaded, after the manner of that which the forefatherd spake.
>
> Copy the fathers, them that have gone before thee _ _ _ _. Behold, their words endure in writing. Own (the book) and read, and copy the knowledge, so that the craftsman too may become a wise man (75–76).

In a society that strove for excellence in all fields of knowledge, it is significant that *Seshat*, the ntr that represented scribes, librarians, writing, and archivists, was considered the foremost builder and was associated with builders very early on: As scholars now confirm: "we can say that building activities and the sacral architecture represent some of the earliest contexts in which *Seshat* appears" (Magdolen 173). It seems that in this cultural context, scribes, librarians, and archivists are considered master builders because it is they who safeguard the secret to building bigger, grander, and perfect structures, passed down knowledge.

So, you see, we have spent centuries admiring the buildings. We photographed them, studied them, and tried to replicate them (unsuccessfully), but we have not expended the same energy studying the belief systems, the codes of conduct, and the ethical principles that are imperative to develop a civilization that can accomplish what they did. Perhaps the secret has been hiding in plain sight. Or maybe the secret would be in plain sight if the library of Alexandria, along with its 490,000 papyri, had not been suspiciously burned down to the ground during the Roman occupation of the country in 48 BC (Thiem 508).

Similarly, the *Teachings of Ptahhotep*, another volume in the wisdom literature (the Sebait) written around 2350–2375 BC, stress the importance and rarity of truly good speech. Again, it is important to note that in the Kemetic context, the adjective "good" really does mean "good" as in accordance with Maatian principles. The fact that later the adjective "good" has come to mean something other than morally righteous is a question best left to linguists. What is important here is to suspend our contemporary understanding of the idea of a good speech being enjoyable, engaging, well organized, extemporaneous, etc., and instead, we reacquaint ourselves with the term in the manner Kemites used it in conversation which is "good" as in ethical.

Ptahhotep is rumored to have been over a hundred years old when he wrote this manuscript as the transmission of *Ka* from father to son (Karenga 179). The contents of his writings are said to be the oldest treatise on morality for which we have evidence (Gray 416). One communication scholar describes the treatise as: "Ptahhotep argued in favour of basic equalities, respect, and the free flow of information and opinions, particularly for political speech, much like social democracy and political liberalism do" (Löwstedt 493). Ptahhotep was not a king, but worked for the King as Vizier (judge, governor, or whatever other services the king required) and thus enjoyed a high level of prestige on Kemetic society during the fifth dynasty.

> Don't be prud of your kenowledge,
> Consult the ignorant and the wise;
> The Limits of aert are not reached,
> No artist's skills are perfect;
> Good speech is more hidden than greenstone,
> Yet may be found among maids at the grindstones (Lichtheim 63)

Both of these excerpts reflect the regard and imperative that Kemites associated with the pursuit of knowledge but what of scientfic knowledge specifically? As discussed in Chapter 1, "The ancient Egyptian science of the stars was prompted in the early eras by the demands of agriculture. Because the harvest seasons and the fertilization of the fields and orchards depended upon the annual inundation of the Nile" (Bunson 67). However, their scientific achievements were not limited for farming technologies and astronomy. The discovery of the *Beatty Papyrus IV*, Chester to the Ramessid Period, a document that dates Nineteenth and Twentieth Dynasties (1307–1070 BCE), has revolutionized our understanding of the Kemetic contribution to the medical sciences:

The papyrus contains medical diagnoses and prescriptions for the treatment of diseases of the anus. The breast, heart, and bladder are also discussed, indicating an advanced knowledge about the human anatomy concerning organ functions and symptoms. Such papyri have offered modern scholars an insight into the sophisticated medical knowledge and practices of the ancient Egyptians, a science that was not attributed to them in the past (Bunson 77).

Perhaps, in looking for evidence of how Maatian ethics promoted the study science, the best place to look is in the all-important Kemetic cosmology. As Bernal states: "The idea that mythology is an allegorical interpretation of historical events or natural phenomena to the masses, who are capable of grasping only a partial truth, was well established in Antiquity" (374).

While this has become a fact to many scholars, it is only recently when a different version in our common intellectual history has come to light: "Egyptian science, despite the tenacious legend to the contrary, was highly theoretical" (Diop 6).

Thus far, the historical record relegated the scientific and mathematical contributions of Kemet to the realm of the practical and it has been the Greeks who have been credited with the theorizing aspects of mathematics. However, it seems that this impression has not been supported by the Greek literature: "According to Democritus and Aristotle, there is no doubt that the Egyptian priests jealously guarded a highly theoretical science behind the thick walls of their temples" (Diop 275).

Perhaps the best evidence we have gathered to confirm Democritus and Aristotles' assertions and understand its connection to Kemetic spirituality and ethics is found in the new evidence regarding the initiative role played by Kemites in the theoretical origin and development of *maathematics*: "The existence of the sacred right-angled triangle shows that, for the Egyptians, some mathematical proportions had a divine essence in the Pythagorean and Platonic sense" (275). Further evidence of the cultural promotion of logical speech can be found in the language of Medu Netcher and the many linguistics tools the Kemites developed in the language to deal with theoretical, scientific, and logical matters.

When Egyptian mathematics, for example, are not smattered or studied superficially, one can find that Egyptian mathematicians dealt rationally with the problems. Indeed, the Egyptians made use of logic as a tool of precision in constructing and developing their mathematics. In geometry – that is, the mathematics of the properties, measurement, and relationships of points, lines, angles, surfaces, and solids or three-dimensional figures – all the problems were arranged in a clear and consistent manner. There is always a logical coherence among the parts of a problem. The basic structure of a problem always consisted of the following parts:

1   *tep*: The Given Problem. This is the precise enunciation of the problem to be solved, with elucidatory examples.
2   *mi djed en. Ek*: Literally, "if one says to you that". This is the stage of definition, where everything is made clear and distinct, and all the relevant terms are explicitly and precisely defined. The expression mi djed means "according to that which is said", that is, the process of reasoning is to be addressed to a precisely formulated problem.
3   *peter or pety*. Literally, this means "What?" In Egyptian grammar, ptr (peter) stands at the beginning of questions with the function of eliciting a logical predicate (Gardiner 1957: 406, §497). A question is an expression of inquiry that invites a reply or solution. At this stage, then, the student is directly required to ponder and analyze (ptr (peter)) the problem under examination.
4   *iret mi kheper*: Correct Procedure. This is the stage of demonstration, that is, the mental process of showing something to be true by reasoning and computation from initial data. The process of calculating is based on a careful set of mathematical formulas.
5   *rekhet. ef pw*: The Solution. This is knowledge (rekhet) found, and grasped in the mind with clarity or certainty. The solution is regarded as true beyond doubt. The student has shown the requisite know-how, that is, the knowledge and skill required to do something correctly. The solution is evident, thanks to the demonstration by a dependable logical procedure.
6   seshemet, seshmet: Examination of the Proof. This is the review of the whole body of evidence or premises and rules that determine the validity of a solution. Such an examination of a logical proof always leads to a further conceptual generalization. Thus, the ancient Egyptians had the technique of forming concepts inductively.
7   *gemi. ek nefer*: Literally, "You have found good". This is the concluding stage. To be able to do something, and find it

correctly done, means that it was done as it should be done. To find (gemi) is to obtain by intellectual effort, and bring oneself to a mental awareness of what is correct, precise, and perfect (nefer). To arrive at a logical conclusion and find that the conclusion withstands critical scrutiny is an achievement in the art of deduction. The adverb nefer ("well") implies that the solution is convincing, so that a contradiction is impossible. The concluding observations are mainly confirmatory. Nevertheless, the rigor of the entire process is evident in the method, and the result is objectively known in all truth (Obenga 41–42).

Par for the cause with the use of logic are the virtues of exercising restraint and good judgment in every day communicative interactions. The following excerpt, from *The Wisdom of Anii* papyrus, is believed to have been produced during the 18th dynasty (1549–1292 BC) and titled "Be cautious in Speech":

Speak not out of thine heart to the . . . man _ _ _ _. A wrong word that "hath come forth from any mouth, if (he?) repreateth it, thou makest enemies (for thyself(. A man falleth to ruin because of his tongue _ _ _ _. A man's belly is broader than a granary, and is full of all manner of answers. Choose thou out the good and speak them, while the bad remain imprisoned in your belly. _ _ _ _

Of a truth thou will ever be with me and answer him that injureth me with falshood, in spite of God who judgeth the righteous. His fate cometh to carry him off (Erman 288).

The fourth and final reason why in the kemetic value system communication was associated with magic stems from the fact that it is through speech that often human beings communicate with the unseen. Whether it is through prayer, commemorating as deceased loved one, or expressing our spirituality, it is often the case that it is through communication that we attempt to bridge a perceived chasm. It is in this way that Kemites felt that there was a magical dimension of communication, as it allowed us to communicate with the unseen. There are, of course, other examples of communication with the dead, i.e. necromancia, but to the degree that a person can take in the advice and wisdom from an ancestor or attempt to communicate with someone whom one cannot otherwise reach, communication does allow for a magical way to feel their presence.

Speaking to Posterity: On the Preeminence of Listening and Silence as Communicative Virtues

The final section of this chapter will combine two signature features of a Maatian approach to ethics: Listening and silence. When contrasted to other classical approaches to ethics and communication ethics, Maat stands heads and shoulders above the rest in its emphasis of these two communication practices as foundational to good communication. In the excerpts that will follow, you will hear about the importance of these two sets of skills in poetry form, stelae (public communication), and in instruction forms. These two aptitudes are widely discussed in Kemetic literature, often together; thus, I will preserve this presentation style in my rendition of them to reader.

The first is an excerpt from public communication. It is an excerpt from the funerary *stelae of Intef, son of Senet* (2065–2000 BCE). The stela is made of limestone and was erected by the son of Intef. Typically, these stone slabs were placed upright and displayed a commemorative message.

> I am knowledge to him who lacks knowledge,
> One who teaches a man what is useful to him.
> I am a straight one on the king's house,
> Who knows what to say on every office.
> I am a listener who listens to the truth,
> Who ponders it in the heart.
> I am one pleasant to his lord's house,
> Who is remembered for his good qualities.
> I am kindly in the offices,
> One who is kind and does not roar,
> I am kindly, not short-tempered,
> One who does not attack a man for a remark,
> I am accurate like the scales,
> Straight and true like Thoth.
> I am firm-footed, well disposed,
> Loyal to him who advanced him.
> I am a knower who taught himself knowledge,
> An advisor, who advice is sought,
> I am a speaker in the hall of justice,
> Skilled in speech in anxious situations (Lichtheim 122).

I've already introduced the *Teachings of Ptahhotep* in the previous section as the author of the first treatise on ethics for which we have

evidence. In the excerpt from his epilogue below, Ptahhotep grounds
the effectiveness of his teachings in his son's (or reader) ability to listen.

> If you listen to my sayings
> All your affairs will go forward;
> In their truth resides their value,
> Their memory goes on in the speech on men,
> Because the worth of their precepts;
> Of every word is carried on,
> They will not perish in this land.
> If advice is given for the good,
> The great will speak accordingly;
> It is teaching a man to speak to posterity,
> He who heard it becomes a master-speaker
> Ir is good to speak to posterity,
> It will listen to it (Lichtheim 73).

I will conclude this chapter with an excerpt from the "The Instruction
Adressed to King Merikare", dating from the second half of the 18th
dynasty. In it, the author illustrates the Kemetic respect and belief in
speech while framing its power in the language of morality. The next
chapter will take a closer look at the language of Medu Netcher as a
vehicle and manifestation of Maatian ethics.

> If you are skilled in speech, you will win,
> The tongue is (a King's) sword;
> Speaking is stronger than all fighting,
> The skillfull is not overcome.
> _ _ _ _ on the mat,
> The wise is a (school) to the nobles,
> Those who know what he knows will not attack him,
> No (crime) occurs when he is near;
> Justice comes to him distilled,
> Shaped in the sayings of the ancestors,
> Copy your fathers, your ancestors,
>
> _ _ _ _
> See, their words endure in books,
> Open, read them, copy their knowledge,
> He who is taught becomes skilled.
> Don't be evil, kindness is good,
> Make your memorial last through love of you,
> Increase the (people), befriend the town,

God will be praised for (your) donations,
One will _ _ _ _
Praise your goddess
Pray for your health --- (Lichtheim 99).

Bibliography

Alkebulan, Adisa A. "The Spiritual Essence of African American Rhetoric." *Understanding African American Rhetoric: Classical Origins to Contemporary Innovations,* edited by Ronald L. Jackson II and Elaine B. Richardson. Routledge, 2002, pp. 23–42.

Assmann, Jan. "Creation through Hieroglyphs: The Cosmic Grammatology of Ancient Egypt." *The Poetics of Grammar and the Metaphysics of Sound and Sign.* vol. 6, edited by S. La Porta and David D. Shulman. Brill, 2007, pp. 17–34.

Ayad, Mariam F. "The Selection and Layout of the Opening of the Mouth Scenes in the Chapel of Amenirdis I at Medinet Habu." *Journal of the American Research Center in Egypt,* vol. 41, 2004, pp. 113–133.

Bunson, Margaret. *Encyclopedia of Ancient Egypt.* Facts on File, 2012.

Dimock, Wai C. "The Planetary Dead: Margaret Fuller, Ancient Egypt, Italian Revolution." *ESQ: A Journal of the American Renaissance,* vol. 50, no. 1, 2004, pp. 23–57.

Diop, Cheikh A. et al. *Civilization or Barbarism: An Authentic Anthropology.* Lawrence Hill Books, 1991.

Erman, Adolf. *Ancient Egyptian Poetry and Prose.* Translated by Aylward M. Blackman. Dover Publications, 1995.

Gray, Giles W. "The "Precepts of Kagemni and Ptah-Hotep"." *Quarterly Journal of Speech,* vol. 32, no. 4, 1946, pp. 446–454.

Karenga, Maulana. *Maat, the Moral Ideal in Ancient Egypt (African Studies).* Taylor and Francis, 2004.

Lichtheim, Miriam. *Ancient Egyptian Literature Volume 1: The Old and Middle Kingdoms.* University of California Press, 2006.

Löwstedt, Anthony. "Do We Still Adhere to the Norms of Ancient Egypt? A Comparison of Ptahhotep's Communication Ethics with Current Regulatory Principles." *International Communication Gazette,* vol. 81, no. 6–8, 2019, pp. 493–517.

Magdolen, Dusan. "A New Investigation of the Symbol of Ancient Egyptian Goddess Seshat." *Asian and African Studies,* vol. 18 no. 2, 2009, pp. 169–189.

Nantambu, Kwame. "Egypt and European Supremacy: A Bibliographic Essay." *A Current Bibliography on African Affairs,* vol. 28, no. 4, 1997, pp. 357–378.

Obenga, Théophile. Egypt: "Ancient History of African Philosophy." *A Companion to African Philosophy.* vol. 28, edited by Kwasi Wiredu et al. Blackwell Publishing, 2006, pp. 31–49.

West, C. S'thembile. "The Goddess Auset: An Ancient Egyptian Spiritual Framework." *Goddesses in World Culture,* edited by Patricia Monaghan. Prager, 2010, pp. 237–248.

6 Medu Netcher
A Picture Says a Thousand Words?

It is estimated that the Medu Netcher language originated 10,000 years ago (Brookfield 102) and it still survives as the liturgical language of the Coptic Church (Allen 2). Kemites did not call their language hieroglyphs, that is a Greek term that translates into "sacred carvings". Instead, Kemites used the term MDW NTR (Mede Netcher) which is translated in various ways: "divine speech", "words of nature", "sacred speech", among others. The Medu Netcher language did not have vowels and thus MDW means "words" or "speech", while NTR, as we have seen throughout this book, is translated as "nature", "spirit", "divine", or "force of nature".

Like any other language, there were various dialects of Medu Netcher spoken in Kemet and the language has undergone a number of changes over its 10,000-year history. Medu Netcher is an ideographical form of communication that has a unique link to communication ethics; this language is not only about speaking the language but about being the language. As "divine speech", Medu Netcher is paradigmatic of the Maatian ethical principles of the Kemetic people.

Medu Netcher has taken many forms throughout its long history and has over 2,000 characters. "The hieroglyphic script is the first Kemetic script first attested in the late fourth millennium. It is made up of phonetic signs for letters, double letters, triple letters, and 'determinatives' which indicate the category of the word's meaning" (Bernal 1010). The second form is hieratic. Kemetic script "gradually developed from Hieroglyphic about 2500 BC. It changed the formal pictorial Hieroglyphic into a cursive script that was still based on the same principles" (Bernal 1010). The third and final form of Kemetic script is demotic. This is a term created by linguists to "classify the Kemetic script, gradually developed from Hieroglyphic about 2500 BC. It changed the formal pictorial Hieroglyphic into a cursive script that was still based on the same principles" (Bernal 1003).

While there is still some debate regarding the origin of the written language (i.e. Mesomotamian or African), scholars have made progress in tracing the history and origin of the Medu Netcher language itself:

> Egyptian civilization is clearly based on the rich Pre-dynastic cultures of Upper Egypt and Nubia, whose African origin is uncontested. Nevertheless, the great extent of Mesopotamian influence, evident from late Pre-dynastic and 1st-dynasty remains, leaves little doubt that the unification and establishment of dynastic Egypt, around 3250 BC, was in some way triggered by developments to the east (Bernal 61).

In addition to a more accurate history of the creation and evolution of Medu Netcher, scholars now have a better understanding of the complexities and idiosyncrasies of the language. The Egyptologists who first encountered and attempted to read Medu Netcher mistook its structure as purely ideographical. An ideograph functions in a language by: "representing ideas (or meaning) with graphic symbols — exposes in the relationship between meanings and graphic symbols. What is called a character in ideographic scripts is actually a syllable" (Muslim 178). Thus, an ideograph is an abstraction of the highest level. "Abstract symbols such as pictographs have a general likeness to the original subject so that viewers can associate them with the real object, person, or environment" (Muslim 178). I will return to this aspect of Medu Netcher, later in the chapter, but for now, it is only important to remember that the structure and function of Medu Netcher is more than ideographic; it is also phonetic and determinative.

Phonetic hieroglyphs represent the sound of the glyph, while determinatives:

> refer to classes of meaning: for example, the sign of the eye refers to everything that has to do with seeing, the sign of the house to all concepts of space, the sign of the sun to concepts of time (Asmann 16).

While a nuanced discussion of the linguistic aspects of Medu Netcher is not warranted here, it is important to discuss the features of ideographic language that enhance communication. One such aspect of ideographs as speech is their ability to reach a much deeper level of consciousness in communication while grounding readers in a particular place. According to one scholar, ideographic language: "Keeps the reader's attention focused as much on the reality signified as on the textual operation: the self-communication of the structure is no longer the primary concern" (Clüver 144).

The use of ideographs is also associated with expanding the creative dimensions of language use as they do more "than communicating a subjective experience or formulating a "message"; they exploit the visual, aural, and semantic qualities of their verbal material and explore the possibilities inherent in the accidents of phonic and visual correspondences (Clüver 137). While at first Egyptologists believed that the Medu Netcher language was purely pictographic, later the discovery of its phonetic and ideographic characters revealed that there is much more to Medu Netcher than meeting the eye:

> The ancient Egyptian hieroglyphs have always been a mysterious writing system as their meaning was completely lost in the 4th century AD. The discovery of the Rosetta stone in 1799 allowed researchers to investigate the hieroglyphs, but it wasn't until 1822 when Jean-Francois Champollion discovered that these hieroglyphs don't resemble a word for each symbol, but each hieroglyph resembles a sound and multiple hieroglyphs form a word (Franken and van Gemert 765).

More research into Medu Netcher has also revealed that far from a primitive, minimalistic language system, it is a highly allegorical form of expression: "The Egyptian priests' use of hieroglyphs was perceived as being linked to their use of allegories and the allegorical significance of the mysteries attributed to them by Plutarch and other Greek writers" (Bernal 321). Whether the language of Medu Netcher was designed to protect Kemetic knowledge or whether it was the product of people who experienced reality in a more heightened and creative way is still being debated.

What we do know is that Medu Netcher was seen by Kemites as a manifestation of Maat. Divine speech was seen as a vehicle for achieving self-actualization and an expression of the sacred. As we will see next, the very structure of Maat works to promote four key Maatian principles: Order, Harmony, Balance, and Truth.

Order

With more than 2,000 characters with phonetic, ideographic, and pictographic functions, Medu Netcher requires optimal organization in order to perform its communicative role. Furthermore, the structure of Medu Netcher was designed to mimic the structure of the landscape that surrounded Kemites. Thus, language and place worked to affirm each other as well as reinforce a priority of values that became more internalized with each articulation and every utterance:

> The most important role that the hieroglyphs played in defining the relation between symbolic expression and the order of the world was in furthering ideas about a natural language. When referring to natural language, modern linguists generally use the term as a synonym for ordinary language; that is, they use it to distinguish between particular instances of language use and the common mental processes that structure language as a whole, or to distinguish between actual spoken languages and any artificial or universal languages (Singer 50).

Because Medu Netcher linguistically anchored itself in the natural, physical features of the landscape, its fauna, and in the many activities that Kemites perform every day to survive and flourish, the expression of the language became naturalized, autochthonous, and an extension of the culture as a whole. Languages with ideographic components tend to reinforce a collective identity since: "Their linguistic value is insured because they serve as symbols for the order of things, and this order presents itself naturally to the mind" (Singer 57).

However, such symbiosis between a native language and its place of origin should not be confused with a primitive or unsophisticated language. Medu Netcher is emblematic of an ancient language that although limited in vocabulary allowed for a creative expression by design. For example, while some scholars have speculated that Medu Netcher did not make use of vowels in order to protect the knowledge accumulated by Kemites for generations, others speculate that the lack of vowels allowed language users to be creative in their choice of sound while conveying a message effectively. Such a practice would be analogous to the consonants of the language functioning as bricks, while the individual choice of mortar material would be considered the individual, creative expression of its users. Additionally, the conventions of the language allowed for other flexible features:

> The reading order in Egyptian hieroglyphs is either from left to right, right to left or from top to bottom. The only indication of the correct reading order is that glyphs will face the beginning of a line, and top to bottom is indicated by columns separators. Multiple horizontal hieroglyphs in a column should be read as a single line (Franken and van Gemert 766).

As we have seen the order, one of the foundational features of a Maatian ethics found easy and complimentary reinforcement in the properties of Medu Netcher. A combination of high-level abstraction and the naturalness of the language promoted a worldview of ordered creativity.

Harmony

According to communication scholar Molefi Kete Asante, the foundation of African philosophy is the principle that: "Life consists in making harmony and peace with nature" (Asante 6). The Medu Netcher language played a part in reflecting and reinforcing this worldview. In particular, its pictographic and ideographic features would have grounded language users in a specific place and time and this allowed them to be more mindful and conscious of the present. As one scholar notes: "Illustrative images, narratives, and moral inscriptions rendered in full spectrum reminded the people of their obligation to each other and to their community". The Medu Netcher (hieroglyphs) inscriptions, images in themselves, instructed people on how to live morally in the moment towards a meaningful and fulfilling future. These sacred images often served as a guide for the deceased, seeking beyond corporeality a spiritual immortality and the grace of the divine (Wilson 578).

> The ideographic features of the language would've kept language users engaged with thoughts, not just words. The difference lies in how uniquely the language grounds users in a particular place and time. For example the use of local fauna, flora, landscapes, and tools both rely on a familiarity with its iconography and foment a deeper engagement and relationship to these images as beyond their phonographic and logographic value, language users would've engaged in a deeper knowledge and appreciation of their attributes, skills, and uses. Such a practice yields ontological that stress inter-dependence and an ecological worldview in which harmonious relationships amongst people and species are privileged and promoted.

Below is a small sample of the characters used in Medu Netcher. Immediately, one is struck by the use of aspects of natural as vehicles for making sense of experience (as language does), to create meaning (as language does), and help negotiate our human relationship to the natural world (as language does). Harmony is a key principle of Maat. In harmonizing with the environment and society around us with do more than fit in or feel good, we promote individual and collective survival by reminding us that we have a non-negotiable relationship of interdependence with the cosmos and all of its substances, including human beings. To harmonize does not mean to lose one self. If one

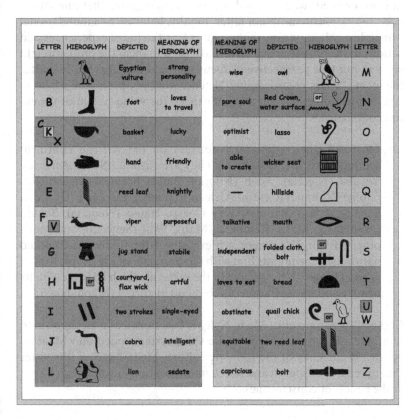

LETTER	HIEROGLYPH	DEPICTED	MEANING OF HIEROGLYPH	MEANING OF HIEROGLYPH	DEPICTED	HIEROGLYPH	LETTER
A		Egyptian vulture	strong personality	wise	owl		M
B		foot	loves to travel	pure soul	Red Crown, water surface	or	N
C K X		basket	lucky	optimist	lasso		O
D		hand	friendly	able to create	wicker seat		P
E		reed leaf	knightly	—	hillside		Q
F V		viper	purposeful	talkative	mouth		R
G		jug stand	stabile	independent	folded cloth, bolt	or	S
H	or	courtyard, flax wick	artful	loves to eat	bread		T
I		two strokes	single-eyed	obstinate	quail chick	or	U W
J		cobra	intelligent	equitable	two reed leaf		Y
L		lion	sedate	capricious	bolt		Z

Figure 6.1 Medu Netcher glyphs.

thinks about it in musical terms, a harmony requires different tones by its very definition. If one tone is only audible, then it ceases to be harmonious and become mono-tone. Medu Netcher, in no way, prescribes monotony; on the contrary, it prescribes unity in diversity.

Balance

The Medu Netcher Symbol for Maat is rectangular and wedge-shaped plinth or base (de Ville 336). Her order was the foundation and the base for the order in the universe. In poetic terms, the ethics of Maat also functioned as the proper foundation to self-actualization. Her ethics provided the stability (and height) that could help one reach the astral world to be close to the sun. Similarly, the ethics of Maat, to do and speak Maat, provided the foundation for a good life, in all of its realms.

To achieve balance, one must pay close attention to how one's actions are affecting the whole and perform corrective action, whether in word or deed, when needed. Balance also requires that one knows one's limitations so as to not falter. Finally, balance, as a foundational principle of Maat, means that we must learn to judge experiences and actions in the proper perspective and assign to them the proper weight so as to not lose perspective or focus on the things that really matter the most. In this regard, Medu Netcher not only promoted Maatian communication ethics but also influenced the value system of the entire continent:

> Most African traditional languages on the African continent are linked to Medu Netcher, They are isiZulu, (South African), Yoruba (Nigeria), Acoli (Nilotic), Banda (Central African Republic), NgBandi (Central African Republic), Luganda (Uganda), Sena (Zambia), Baoule (Ivory Coast), Hausa (Nogeria), Senufo (Manianka, Mali), Bambara (Mali), Wolf (Senegal), and many other African Languages... It explains **nature** to him as could no naturalist; it acquaints him with the character of the people wound around him with the society in the midst of which he lives, with its history and its aspirations as could no historian (Maphalala 373).

In short, the language of Medu Netcher functioned as a mechanism to reinforce and advance the Kemetic value system coded into Maatian ethics. As currency of the Kemetic culture, Medu Netcher illustrates the supremacy of speech and how communication practices can be put at the behest of consciousness.

Transcendental Truths

The final aspect of Medu Netcher that promotes Maatian ethics is its focus on truth. For Kemites, anything that is real is eternal. In other words, there was a definitive transcendental aspect to Kemetic spirituality and ethics premised on the idea that "the good" is that which transcends time. Please note that I did not say place and time. There is a reason for that; Kemitic culture was not focused on escaping mortality. On the contrary, the blessed, the "true of voice" who would enter the future life, would spend it, in this belief system, cultivating the earth for the benefit of promoting all life on the planet. In other words, in the best possible scenario, getting into the future life, one would stay right here contributing to the earth in the "fields of the blessed".

Time was another matter. To be remembered by one's community, especially after moving on to the future life, was of paramount importance. It is no wonder that writing was so vital to people who saw it as their duty to learn from the past and teach to the future. Medu Netcher, as a result, was an instrument of said transcendence as the written script is one way to reach across time and through the magic of communication reach future generations. However, Medu Netcher also served a reflective and edifying function of internal communication (monologue) as the language itself worked to reinforce values and negotiate the individual relationship to the collective and the universal.

This focus on transcendence does have similarities with the deontological view of ethics as well as the Kiekegaardian view of the role of monologue in moral development. For even if we grant that a certain level of abstraction is necessary and inherent to ethical communication, we still have not fully explored the possibility that what is now commonly referred to as a monologic act, such as the categorical imperative, or speaking to the ancestors, should be considered a form of communication proper. First, the aesthetic components of Kant's theory of judgment are relevant in order to achieve a more complete description of the "enlightenment individual".

In order to assess the strengths and weaknesses of Maat as an approach to communication ethics as well as its viability, we must make it conversant with classical and contemporary ethical theories. This recovery effort far from aiming to defend ethical models based on non-participation is an attempt to determine whether, in fact, the harsh criticisms that have been made of transcendental moral theories have considered the full range of concerns and communication types that some of these models, including Maat, were attempting to address. Nor I am suggesting that what has been described here thus far as internal communication is in anyway a substitute for interpersonal ethical communication; instead, I have tried to argue that both internal and external (interpersonal) types of communication may be considered as complimentary to one another and applicable to ethical discourse.

Of course, it is important to note that while terms found in Maat and Kant such as a-priori and transcendental seem to disregard or displace human experience, or reduce them to something of lesser importance, once we accept the basic distinction between phenomena and noumena, or the "ba" and the "ka", it becomes more difficult to read too much into some of the terminology used to describe some of the basic organizing mechanisms of human experience. For example, although it may seem that when Kant speaks about a-priori principles as guiding possible experience, he is implying that our experience of the world is inferior to these abstract organizing principles, but he is not. For Kant, our possible experience of the world is the supreme knowledge available to us since the noumenal is foreclosed to us. Hence, an organizing principle of human experience, like the Ka, may seem at first detached from that experience, indeed, before or prior, to our experience, but in Kant's case, we must be careful not to interpret "prior" as anything more than the way in which it is organized in order to be cognized or experienced.

Our cognitive apparatus, for example, way in which we are able to perceive a tree (through its color, shape), may be wired into us but there are still significant individual differences in perception. One must not confuse the general mechanisms through which we, as human beings, are able to experience that tree with the idea that these properties (of color and shape) are what is really true about them or that the shape and color of the tree exist independent of our experience of the tree and our human understanding of it. For Kant, there is no way we can know what is really true about "things-in-themselves", so our experience of the world is all we have to rely on. The view that is often attributed to Kant is more reminiscent of a Platonic theory of forms than a Kantian view of a-priori principles.

Transcendental ideas and a-priori concepts, then, while able in their own function to shape (again organize) our perception of the world of possible experience, do not constitute its ultimate reality and Kant certainly does not suggest that they arise ex-nihilo. For Kant, we are hopelessly embedded in our limited way of "knowing" the world, and while these limitations may at first appear to apply only to our cognitive faculties, the idea that there are limits to our certitude and the idea that everything that pertains to human existence is rooted in our experience of the world extend through all of Kant's thought. There is a difference between the available ways in which we are able to cognize and "know" our world and the way it really is outside of our perception and our experience of it. But this distinction may still leave us to wonder how abstraction, especially in a Maatian sense, is a necessary feature of ethical theory.

Let's start with Kant. For him, reason is the "faculty of syllogistic reasoning by which one draws an inference" (Caygill 347). Notice the way in which rather than dealing with certainty of facts, we are presented in this definition with words like "syllogistic" which he borrows from Aristotle, dealing with "wordly facts" that do not exclude material reality and the "inferences" as delimiting the function and scope of reason. At the center of this type of deliberation for Kant lies the relationship between the individual and her experience of the world. Now, rather than suggesting that having the same cognitive mechanisms and having, as humans, to rely on reason for ethical discernment, Kant insists that valid ethical rules and standards cannot come from outside a person, but have to originate from a person's own active mind (will). Here, we can draw a parallel between the term "autonomous" and this idea of Kant that "ethical rules and standards" have to originate from a person's own active mind. For I still want to insist on a reading on Kantian autonomy that does not preclude the possibility that through reflective judgment and what Kierkegaard might call indirect communication, a person is able to internalize and align himself with these rules, after having participated vigorously in the evaluation of these rules through dialogic interaction with others. I don't see why the two modes of ethical communication must be mutually exclusive.

Thus, if we grant that both communication processes are possible and not necessarily in opposition with one another, it seems to be the case that we need to envision at least two different communication processes in order to address at least two different types of ethical modalities. One modality has to do with the level of certainty that can be achieved via the communication of more objective, verifiable information, like the scientific pursuits of the Kemites, and another, the internal, a mode of communication, that does not require a physical interlocutor, yet allows individuals to make her own thoughts the object of inquiry.

It is my contention that not only is ethical monologue possible, and common, but it is one of the most important communications one can have in the realm of ethics. For unless we are able to convince ourselves that a particular ethical principle is worth upholding, that we want to become "true of voice" for example, and that we are able to do so with the confidence that we are not under some type of external coercion, force, or mind-control, then ethical principles cannot be fully considered democratic and we run the risk of falling into what Cornell West has described as a world in which even something like truth becomes ideological.

Finally, from the first two critiques, we know that Kant was very much preoccupied with the idea of how information becomes knowledge, how we collect information through our bodies, and how we judge things in a proper manner. In his discussion of judgments, he points out that although the study of this part of the cognitive process is essential, aesthetic judgments, for example, do not ultimately have any bearing on how well we know the representation of an object. For Kant, there are three ultimate modes of consciousness: Knowledge, feeling, and desire (13). So why would feeling be considered the only mode of consciousness that does not enhance our knowledge of an object or its representation? The answer seems to be that judgment corresponds to the feeling of pleasure and pain and "pleasure is inter-mediate between our perception of an object and our desire to possess it" (Critique of Judgement xvi). However, because aesthetic judgments are reflective of our pleasure in representation (not based on con-cepts), no knowledge of an object can be derived from them. How-ever, through aesthetic judgments, there is plenty to find out about ourselves. For example, the Medu Netcher pictographs remind us of place and our relationship to the earth. Isn't this information also part of who we are fundamentally? Part of our truth? The aesthetic dimen-sion of Maatian ethics says a resounding yes.

The will to truth, which, in Nietzschean terms, could be described as a will to power (75), is not something that belongs to each of us uniquely, but a drive which is a part of the world we inhabit. Therefore, it is up to us, to each of us to cultivate internal reflection for our sake and the sake of future generations. Nietzsche states, "what is great in man is that he is a bridge and not and end: what can be loved in man is that he is an overture and a going under" (127).

Maatian ethics call for interpersonal and internal engagement for the sake of willfully pursuing the goal of becoming part of a universal consciousness. G.W.F. Hegel and Brazilian philosopher of education Paolo Freire share some similarities with Maat in this respect. Hegel and Freire both theorize about the relationship between reflection, consciousness, and praxis. Both thinkers argue that a meaningful awareness about one's relationship with the self and others can lead to effectively transforming our actions into guided practice. For He-gel, consciousness and self-consciousness are necessary moments of awareness of *Geist* in our trajectory to absolute spirituality. For when we become conscious, we become engaged in contemplation and con-templation is the first step toward transcending our natural state. He-gel notes that, in a natural stage, we are immersed in negativity, as we contemplate things as if they are separate and different from us. In a state of nature, we have not quite figured out how is it that we are related to the things we observe. We look at the world around us as if it had a separate truth or essence that we can get at through sense perception and through our observation (146).

Freire's "Pedagogy of the Oppressed" is anchored on this idea of consciousness and at least some degree of internal reflection too. When Freire talks about *conscientizacao*, which is the crowning jewel of Freire's pedagogical theory, he means "indirectly" (maieutically) making individuals aware of their place in the world. The poor, he says, must become conscious of their own history through a deeper, more daring understanding of the world (Freire 74). Through reflection, dialogue, and an ability to read and write, the poor can come to realize the unjust state of their lives and begin to envision how to transform their reality of oppression. In other words, the poor can achieve consciousness by developing an awareness of the power relations of which they are partaking, willingly or not. In that way, education can function as an "engine" for revolution. But notice that while education can function as an engine for revolution without reflection, without the individual's ability to come to this realization, at least partly, on their own, this education, he is describing for us, would become little more than some (well-meaning) form of mind-control. And the kind of revolution that Freire is prescribing has the self deeply situated and aware of their own consciousness before this revolution he speaks of can take place. Therefore, as Foucault argues, "It is the power over self which will regulate the power over others" (8).

A Maatian approach to ethics differs from other classical approaches in that it views aesthetic feelings as communicable, in fact, universally communicable. On the contrary, Kant claims that these judgments of taste, as such, cannot lay a claim of universal validity unless, of course, we are dealing with a judgment which is a-priori and not empirical. For Kant, aesthetic judgments have some a-priori basis, in so far as they have principles and rules that allow us to organize the perception as phenomena. However, when he describes objects that are purely beautiful, like a peacock feather, he refers to objects, or representations of objects which are aesthetically judged solely based on their form and which are not corrupted by charm or any other matter of personal taste.

Kant explains that although a judgment of taste justly claims the agreement of all men "because the ground of this pleasure is bound in the universal" (Critique of Judgement 28), these judgments are subjective and singular and can only be made by experiment. Part of the reason for this is that the pleasure that characterizes aesthetic judgments is bound up with representation. Therefore, unless we empirically encounter an object for which a representation is produced, there can't be pleasure.

What seems to be at stake here is the idea that when we talk to each other about what is beautiful, or when more than one person comes into contact with the same beautiful object (or its form to be more precise), they have, in fact, an ability to share and maybe agree upon what kind of feeling this form of representation of this object has provoked in them. This way, such an exchange cannot be said to be based on the idea of any concept, since feelings for Kant are (1) not representable (they just are) and (2) the exchange cannot be said to require agreement based on a concept, since feelings are not in the realm of logic or reason. This idea of *free play* then is what ultimately holds this idea of universal communicability together, since feelings (or mental states) are not limited or shaped by any concept, so a mutual agreement can be achieved that the feelings being expressed are indeed mutual.

It is an interesting fact that among the 2,000 characters of Medu Netcher language and all of its possible combinations, "There was no word in the language (Medu Netcher) for Spirituality, perhaps because all aspects of life were 'Spiritual'" (115). Instead, the language was based on carefully chosen symbols that offered a road map to the virtuous qualities of a powerful communicator, one of which was silence. Silence, after all, allows us to listen carefully, to be introspective and according to Ptahhotep silence also allows one to express communicative excellence by either showing deference, listening to (an)other or exercising much needed restraint. As the Medu Netcher illustrates, words evoke power, a sense of place, a moral compass, but so does silence.

Bibliography

Allen, James P. *Middle Egyptian; an Introduction to the Language and Culture of Hieroglyphs.* Cambridge University Press, 2010.

Amen, Rhkty W. "The Philosophy of Kemetic Spirituality." *Reconstructing Kemetic Culture,* edited by Maulana Karenga. University of Sankore Press, 1990, pp. 115–130.

Asante, Molefi K. *The Egyptian Philosophers: Ancient African Voices from Imhotep to Akhenaten.* African American Images, 2000.

Assmann, Jan. "Ancient Egypt and the Materiality of the Sign." *Materialities of Communication,* edited by Hans Ulrich Gumbrecht and K. Ludwig Pfeiffer. Stanford University Press, 1884, pp. 15–31.

Bernal, Marin. *Black Athena: The Afroasiatic Roots of Classical Civilization.* vol. 1. Rutgers University Press, 1987.

Brookfield, M. "The Desertification of the Egyptian Sahara during the Holocene (the Last 10,000 years) and Its Influence on the Rise of Egyptian Civilization." *Landscapes and Societies,* edited by I. Martini and W. Chesworth. Springer, 2011, pp. 91–108.

Caygill, Howard. *A Kant Dictionary.* Blackwell Publishing, 1995.

Clüver, Claus. "Reflections on Verbivocovisual Ideograms." *Poetics Today,* vol. 3, no. 3, 1982, pp. 137–148.

de Ville, Jacques. "Mythology and the Images of Justice." *Law & Literature,* vol. 23, no. 3, 2011, pp. 324–364.

Franken, Morris and Jan van Gemert. *Automatic Egyptian Hieroglyph Recognition by Retrieving Images as Texts.* ACM, 2013.

Foucault, Michel. *The Final Foucault.* Edited by James Bernauer and David Rasmussen. MIT Press, 1988.

Freire, Paulo. *Pedagogy of the Oppressed.* Continuum, 2000.

Gardiner, Alan H. "The Nature and Development of the Egyptian Hieroglyphic Writing." *The Journal of Egyptian Archaeology,* vol. 2, no. 2, 1915, pp. 61–75.

Maphalala, J. S. "African Languages: Obstacles to Internationalism or Additional Wealth for the World?" *Les Historiens Africaines et la Mondalisation* Actes du 3e congrès international, edited by Issiaka Mandé and Blandine Stefanson, Ashima, 2005, pp. 363–378.

Muslim, Zulkifli. "Ideographic Symbols, Meanings in Visual Communication Design. International Journal of Arts and Humanities, vol. 2, no. 4, 2018, pp. 278–290.

Nietzsche, Friedrich. *The Portable Nietzsche.* Edited and translated by Walter Kauffman. Penguin Books, 1976.

Nussbaum, Martha C. *Women and Human Development: The Capabilities Approach.* Cambridge University Press, 2000.

Singer, Thomas C. "Hieroglyphs, Real Characters, and the Idea of Natural Language in English Seventeenth-Century Thought." *Journal of the History of Ideas,* vol. 50, no. 1, 1989, pp. 49–70.

Wilson, Khonsura A. "Imaging and Imagining the Cosmos: A Creative Ideal and Meme Defined by Form, Feeling, and Function." *Journal of Black Studies,* vol. 42, no. 4, 2011, pp. 577–592.

7 A Lust for Life
Allegory and Poetry

An unfortunate byproduct of the egyptomiac narrative of Kemet that has dominated the popular imagination for decades is the perception of Kemet as a death-cult society. Depictions of ancient Africans, including Kemites, as obsessed with death have concealed the much richer Kemetic arts and expressions of love, joy, exhilaration, heartbreak, and the full range of human emotions Kemites not only experienced but left written for posterity. Lost in the zeal to Other, mythicize, and vilify Kemetic culture are the voices of Kemites themselves who expressed their joy in partaking Maat through short stories, poems, anecdotes, biographies, and many other written artforms. Scholars have made efforts to dispel this misperception by offering more nuanced and accurate accounts of cultural attitudes of death in the African context but the stereotypes and caricatures continue. As one scholar laments:

> Engagement with death is not only philosophical but also psychological. This African attitude to death has misled some to assume, as Russel (2004) does with particular reference to the ancient Egyptians, that the Egyptians were preoccupied with death. To the contrary, Egyptians had an obsession with eternal life (Fletcher, 2015). As Carruthers (1984) notes, it must be emphasised that ancient Egyptians were neither obsessed nor preoccupied with death. To the contrary, they enjoyed life and were quite realistic about death (ibid). With this realisation all or most of what they did was with the future destination in mind. Keeping death in mind at all times was/is a reminder to prepare for the ultimate - a departure to the abode of the ancestors (Sesanti 67).

While it is impossible to offer an exhaustive selection from the personal writings of Kemites, I wish to allow the final chapter in this book to be that of the voice of the ancestors themselves, telling their stories in their own voice. For the purposes of organization, I have selected pieces that represent the sort of concerns moral agents spend much of their time contemplating: Love, family, sorrow, friends, and nature. My hope is to allow these voices to speak to the reader without filter (other than the translation which is unquestionably significant), across time, just as they hoped.

An ethical life is the one that can offer feelings of satisfaction and happiness as well as serious reflection and conflict. I will end this brief exploration of the moral ideal of Maat in a communication context with aspects of the ancient Kemetic life seldom seen, heard, and acknowledged. It is they who will finish telling this story. In the spirit of Ankh Mi Ra, let's do our part to remove the mysticism from Kemetic life, and "Let the ancestors speak!".

Maat in the Ancestors' Voice

Kemites often acknowledged and referenced Maat in public and private writings. The small selection below illustrates the Kemite view of speech as a central concern of ethics. Maat, as the ntr of truth, was seen as the moral ideal for human communication. The first piece is an excerpt from a *Prayer and a Hymn of General Haremhab*, a general who became Pharaoh and reigned between 1323 and 1309 BC. The second piece is an excerpt from the *Dedication inscriptions of Seti I* located In the Rock Temple of Wadi Mia. The inscriptions are believed to be from the New Kingdom (1550–712 BC). The final excerpt is from *The Tale of the Eloquent Peasant*, a rich and cleverly written treatise on morality from the Middle Kingdom (2030–1650 BC).

(1)
A royal offering to Thoth, lord of writing, Lord of Khmun
Who determines Maat, who embarks Re in the night-bark,
May you let the speech be answered for its rightness.
I am a righteous one toward the courtiers,
If a wrong is told me,
My tongue is skilled to set it right.
I am the recorder of royal laws,
Who gives directions to the courtiers,
Wise in speech, there is nothing I ignore.
I am the adviser of everyone,
Who teaches each man his course,

Without forgetting my charge.
I am the one who reports to the Lord of the Two Lands.
Who speaks of whatever was forgotten,
Who does not ignore the words of the Lord.
I am the herald of the council,
Who does not ignore the plans of his majesty;
For the ka of the Prince, Royal Scribe, Haremhab, justified.

(Lichtheim 101)

(2)
Do not rejoice ------. He who damages the work of another,
the like is done to him in the end. A despoiler's monuments
are despoiled. A liar's deed does not endure. A king's
[strength] is *[maat]*. [Listen to me, ye kings who shall be
after me] --- to let you know; I foretell from afar so as to
protect you.

(Lichtheim 66)

(3)
Do not speak a falsehood for you are great.
Do not be a lightweight for you are weighty.
Do not speak a falsehood for you are a balance.
Do not act confused for you are the chief.
Look, you are the main one like the balance.
If it tilts, you may tilt.

Do not stray; take the oar; pull on the tiller rope.
Do not rob, take action against the robber.
Not great is the great one who is covetous.
Your tongue is the plummet;
Your heart is the weight;
Your lips are its arms.

(Fisher 21)

Love

The small selection below offers a small window into the exuberance
and joy of Kemetic poetry. At times, it is playful, impassioned, funny,
and all the things we associate with romantic love. They also illustrate
the role family and morality played in the possibility for such relationships. The first excerpt is from the love songs of *Papyrus Harris 500*

from the New Kingdom (1550–712 BC). The second excerpt is a song, one of the twenty *Cairo Lone Songs* also from the New Kingdom. The final excerpt, also from the New Kingdom, is from one of *The Love Songs of Papyrus Chester Beatty*.

(1)
(How) intoxicating are the plants of my garden!
[The lips] of my beloved are the bud of a lotus,
Her breasts are mandrakes,
And her arms are ornate [...].
Behold, her forehead is a snare of willow, And I am a goose.
My [hands are in] her hair as a lure,
Held fast in the snare of willow.

(Simpson and Ritner 309)

(2)
When I embrace her, her arms open wide before me
As if for one who (has returned) from Punt.
It is like a misty plant blooming in its fullness,
Whose perfume is that of laudanum.
Her lips open wide as I kiss her,
And I rejoice (even) without beer.

(Simpson and Ritner 318)

(3)
Wise is my mother in commanding me:
"Give up looking after such things!"
Behold, my heart is tormented when it remembers him,
For love of him takes hold of me.
Behold, he is senseless,
But yet I am exactly like him.
He knows nothing of my desire to embrace him
Or that he should contact my mother.
Oh my beloved!
I have been fated for you By the Golden Goddess of women.
Come to me, that I may see your beauty,
And my father and mother will be happy.
All my people will rejoice together because of you,
They will rejoice because of you, my beloved.

(Simpson and Ritner 324)

Family

As a communitarian ethics, kinship with one's family was an important dimension of Maatian ethics. The first excerpt is from the *Instructions of Ptahhotep*, written in the Old Kingdom (2575–2150 BC). The two other excerpts are from the New Kingdom's *The Tale of the Eloquent Peasant* and illustrate how a Maatian ethics extended the concept of family beyond the narrow scope of blood relationships.

(1)
Do not be selfish with respect to your relatives,
For greater is the claim of / the good-natured man than that of the assertive.
He who forsakes his relatives is (truly) poor,
For he lacks the compassion to respond to their entreaties.
Even a little of what one yearns for Can calm a distressed man.
If you are well-to-do and establish your household,
Be gracious to your wife in accordance with what is fair.
Feed her well, put clothes on her back;
Ointment is the balm for her body. Rejoice her heart all the days of your life, For she is a profitable field for her lord. Do not condemn her.

(Simpson and Ritner 139)

When you go down to the Lake of Ma'at, 33
You will sail on it with a good wind.
No part of your sail will be torn,
Nor will your ship stall.
No disaster will befall your mast
Nor break your yards.
You will not be too powerful,
Nor will you run aground. (B1 90)
You will not be carried away by a wave;
You will not taste the evil of the river.
You will not behold fear.34
Darting fish will come to you.
You will catch fat birds.
For you are a father to the orphan,
A husband to the widow,
A brother to the divorced.

(Fisher 16)

Double the food your mother gave you, Support her as she supported you;

She had a heavy load in you,
But she did not abandon you.
When you were born after your months, She was yet yoked <to you),
Her breast in your mouth for three years.
As you grew and your excrement disgusted,
She was not disgusted, saying: "What shall I do!" When she sent
you to school,
And you were taught to write,
She kept watching over you daily,
With bread and beer in her house.
When as a youth you take a wife, And you are settled in your
house, Pay attention to your offspring, Bring him up as did your
mother.
Do not give her cause to blame you, Lest she raise her hands to
god, And he hears her cries.

(Lichtheim 145)

Sorrow

Sadness and despair are part of the human experience in any culture.
The small selection included here showcases the existential, romantic,
and spiritual crises and offers some insight into the kinds of things
Kemites contemplated in times of hardship. The first excerpt is from
the *Lamentations of Khakheperre-Sonb*, estimated to have been written during the 18th Dynasty of the Middle Kingdom. It speaks eloquently of the healing aspects of communication, especially during
trying times and illustrates the Maatian virtue of restraint. The remaining two excerpts are from *The Man Who Was Weary of Life*, also
from the Middle Kingdom.

(1)
My malady is long and heavy,
The sufferer lacks strength to save himself
From that which overwhelms him.
It is pain to be silent to what one hears,
It is futile to answer the ignorant,
To reject a speech makes enmity;
The heart does not accept the truth,
One cannot bear a statement of fact,
A man loves only his own words.
Everyone builds on crookedness,
Right-speaking is abandoned,

I spoke to you, my heat, answer you me,
A heart addressed must not be silent,
Lo, master and servant fare alike,
There is much that weighs upon you!

(Lichtheim 122)

(2)
"I opened my mouth in response to my ba, answering what he
had said:

'This is become too onerous for me today;
my ba is not in accord with me.
This is even worse than opposing me; it is like forsaking me!
But my ba shall not depart! 1
He must stand now as my defense in this. (I shall restrain him)
in my body as with a net of rope.
Never shall he succeed in fleeing on the day of anguish.

But behold! My ba would deceive me,
but I heed him not,
While I am impelled toward a death whose time has not yet come.
He flings me on the fire to torment me […]

And yet he shall be within me on the day of anguish;
He shall stand (with me) in the West
as one who perfects my happiness.
Though he would now depart, yet shall he return.
My ba is senseless in disparaging the agony in life
And impels me to death before my time.
And yet the West will be pleasant for me,
for there is no sorrow there.
Such is the course of life, and even trees must fall.
So trample down my illusions, for my distress is endless!

Thoth will judge me, he who satisfies the gods,
Khons will defend me, / he who records the truth.
Re will hear my words, he who guides the sun barque,
Isdes will defend me in the sacred hall,
For my longing is too intense to bring me any joy,
And only the gods will purge my innermost pain."
What my ba said to me: "Are you not a man?
At least you are alive!

So what do you gain by pondering on your life like the owner of
a tomb,
One who speaks to him who passes by about his life on earth?
Indeed, you are just drifting; you are not in control of
yourself,
And any rogue could say, 'I shall guide you.'
You are, in effect, dead, although your name still lives.
The beyond is the place of rest, the desire of the heart;
he West is the (final) landing place,
But your journey (has not yet reached its end)."

(Simpson and Ritner 179–181)

(3)
Whom can I trust today?
One's brothers have become evil,
And friends of today have no compassion.

Whom can I trust today?
Hearts are greedy,
And every man steals his neighbor's goods.

(Whom can I trust today?)
Compassion has perished,
And violence attacks everyone.

Whom can I trust today?
(Men) are pleased with the evil
Which everywhere throws goodness underfoot.

Whom can I trust / today?
Though a man be woeful through ill fortune,
His evil plight causes all to mock him.

Whom can I trust today?
Men plunder,
And everyone robs his comrade.

Whom can I trust today?
A reprobate is my closest friend,
And the companion with whom I associated has become a foe.

Whom can I trust today?
There is no remembrance of the past,
And men now do not treat one in accordance with one's deeds.

Whom can I trust today?
One's brothers have become evil,
And one turns to strangers for integrity.

Whom can I trust today?
People are indifferent,
And every man is sullen to his comrades.

Whom can I trust today?
Hearts have become greedy,
And no man has a heart which can be trusted.

Whom can I trust today?
There are no righteous men,
And the land is abandoned over to the lawless.

Whom can I trust today?
There is emptiness in faithful friends,
And one must turn to strangers for comfort.

Whom can I trust today?
None are contented,
And he with whom one walked is now no more.

Whom can I trust today?
I am laden down with sorrow,
And there is none to comfort me.

Whom can I trust today?
Evil runs rampant throughout the land, Endless, endless evil.
                                 (Simpson and Ritner 184–185)

Friends

Most of us have experienced the joys and difficulties associated with navigating friendships. Kemites are no different in this respect. The excerpts below Kemites will speak on the value and politics of friendship. The first and third excerpts are from the Old Kingdom's *Instructions of Ptahhotep*; the second excerpt is from the New Kingdom's *The Tale of the Eloquent Peasant*.

(1)
Gratify your friends with what has come into your possession,
For what has come to you is a boon from God.
As for him who fails to gratify his friends,

People will say that he is a selfish individual.
No one knows what will come to pass when he considers tomorrow,
And the righteous individual is he by whom men are sustained.
If deeds deserving of praise are done,
One's friends say, 'Well done!'
One cannot bring satisfaction to an (entire) town,
But one can bring happiness (to) friends when there is need.

(Simpson and Ritner 159)

(2)
He who looks too far ahead will become disquieted,
So do not dwell on what has not yet befallen,
And do not rejoice about what has not yet happened.

Patience prolongs friendship,
But as for him who neglects a fault which has been committed,
There is no one who knows what is in his heart.
If law is subverted and integrity destroyed,
There is no poor man who will be able to live,
For he will be cheated, and Ma'at will not support him.

(Simpson and Ritner 40)

(3)
If you desire that friendship should endure In a house which you enter A
s a lord, as a brother, or as / a friend:
In any place which you enter,
Avoid approaching the women,
For there is nothing good in any situation where such is done.

It is never prudent to become overly familiar with them,
For countless men have thus been diverted from their own best interests.
One may be deceived by an exquisite body,
But then it (suddenly) turns to misery. (All it takes is) a trifling moment like a dream,
And one comes to destruction through having known them.

Pricking the jealousy of a rival is a nasty piece of business;
A man may perish because of so doing, if the heart becomes ensnared.

As for him who is ruined through becoming embroiled with
them.
No venture will ever be successful in his hand.

(Simpson and Ritner 138)

Nature

At the core of a Maatian ethics lies a deep appreciation of nature and
our interdependence with it. The language of Medu Netcher itself
was built on allusions to the natural world and thus Kemetic literature exhibits an intimate and joyful view of the human relationship to
nature. The first excerpt *is* a playful story where the trees in the garden are anthropomophized to interact with its human neighbors. *In
this story the trees narrate events and their emotions take center stage.*
The title of the story, dated to either the Middle or New Kingdom,
is *The Trees in the Garden*. The second piece is the longest of the
illustrations and there is a good reason for that. The New Kingdom's
Great Hymn to the Aten illustrates like no other the Kemetic sense
of place and the spiritual connection they felt to it. Each stanza of
this hymn dramatizes the highly localized aspects of the Maatian
universal approach to ethics. The final excerpt is from *The Eloquent
Peasant* and it communicates in a powerful way, the way Kemites
related to the earth, to soil as both the origin of human life and its
inevitable destination.

(1)

_ _ _ _ The . . . –tree speaketh : My stones are like unto her teeth,
and my shape unto her breasts. (I am the best) of the orchard, I
abide at every season, that the sister may recline (beneath me?)
with her brother, when they are drubken with wine and shedeh,
and besrinkled with kemi-oil. _ _ _ _ All (trees) in the garden save
me fade away; I endure twelve months _ _ _ _I stand . . ., and if the
blossom fallrth off, that of the year before is still upon me.

I am the first of all trees and I wil not that I should be regarded
as second. If this is done again, I will no longer keep silence and
betray them, that the wrong-doing may be seen and the beloved
be chastised, that she may not _ _ _ _. The poet then speaks of the
feast with its lotus flowers, blossoms abd buds, its unguents and
beer of all kinds, that she may vasuse thee to pass the day in merriment. The booth of rushes is a sheltered spot _ _ _ _

I see him, he is really coming. Let us go and flatter him. May he
pass the whole day _ _ _ _.

The fig-tree moveth its mouth, and its foliage (?) cometh and saith: _ _ _ _ to the mistress.

Was there ever a lady like me (Yet) if thou hast no slave I will be thy servant. I was brought from the land of _ _ _ _ as a spoil for the beloved. She had me set in the orchard, she putteth not for me _ _ _ _. I (busy?) myself with drinking. and my belly hath not become full of well-water. I am found for pleasure, _ _ _ _ one that drinketh not. By my ka! O beloved, _ _ _ _ bringeth me into thy presence.

The little sycamore, which she hasth planted with her hand, it moveth its mouth to speak. The Whispering (?) of the leaves is as sweet as refined honey. How charming are its pretty branches, verdant as . . . It is laden with neku-fruits, that are reder than jasper. Its leaves are like unto malachite, and are . . . a glass. Its wood in its colour like unto neshmetstone and is . . . as the besbes tree. It draweth them that be not (already) under it, its shadow is so cool.

It slippeth a letter into the hand of a little maid, the daughter of its cheirf gardener, and maketh her run to the beloved : "Come, and pass the time in the midst of thy maiden (?). The garden is in its day. There are bowers and shelters (?) there for thee. My gardeners are glad and rejoyce when they see thee. Send thy slaves akhead of thee, supplied with their utensils. (Of a truth) one is (already) drunken when one hasteneth to thee. Were one hath yet drunken.(But) the servants come from thee with their vessels, and bring beer of every sort, and all manner of mixed loaves, and maby flowers of yesterday afeast is nd to-day, and all manner of refreshing fruit.

"Come, and spend the day in merriment, and to-morrow, and the day after, three whole days, and sit n my shadow."Her lover siteth on her right hand. She math him drunken amd heedeth all that he saith. The feast is diordered with drunkeness, and she stayeth on with her brother,

Her . . . is spread out under me, when the sister walketh about. But I am discreet and speak not of what I see. I will say no word. (Erman 249–251.)

(2)
Worshipping (The Living One) Re-Horakhty who Rejoices in the Horizon) | (In his Identity as the Light who is in the Aten) | living forever and ever, the Living Aten, the Great One who is in Jubilee, Master of all that the Aten encircles, Master of Heaven, Master of the Earth, Master of the Per-Aten in Akhet-Aten; 1 and

the King of Upper and Lower Egypt, the one Living on Maat,
Lord of the Two Lands (Nefer-kheperu-Re Wa-en-Re) |, son of Re,
living on Maat, Master of Regalia (Akhenaten) |, the long lived;
and the Foremost Wife of the King, whom he loves, the Mistress
of Two Lands (Nefer-nefru-Aten Nefertiti) |, living, well, and
young forever and ever.

He says:

You rise in perfection on the horizon of the sky,
living Aten, who determines life.
Whenever you are risen upon the eastern horizon
you fill every land with your perfection.
You are appealing, great, sparkling, high over every land;
your rays embrace the lands as far as everything you have made.

Since you are Re, you reach as far as they do,
and you curb them for your beloved son.
Although you are far away, your rays are upon the land;
you are in their faces, yet your departure is not observed.

Whenever you set on the western horizon,
the land is in darkness in the manner of death.

They sleep in a bedroom with heads under the covers,
and one eye cannot see another.
If all their possessions which are under their heads were stolen,
they would not realize it.
Every lion comes out of his cave and all the serpents bite,
for darkness is a blanket.
The land is silent now, because He who makes them is at rest on
His horizon.

But when day breaks you are risen upon the horizon,
and you shine in the Aten in the daytime.
When you dispel darkness and you give forth your rays
the two lands are in a festival of light,
alert and standing on their feet,
now that you have raised them up.
Their bodies are clean, and their clothes put on;
their arms are » lifted ... in praise at your rising.

The entire land performs its work:
all the flocks are content with their fodder,

trees and plants grow,
birds fly up to their nests,
their wings » extended ... in praise for your Ka.
All the kine prance on their feet;
everything which flies up and alights,

they live when you rise for them.
The barges sail upstream and downstream too,
for every way is open at your rising.
The fishes in the river leap before your face
when your rays are inside the sea.

You who places seed in woman
and makes sperm into man,
who brings to life the son in the womb of his mother,
who quiets him by ending his crying;
you nurse in the womb,
giving breath to nourish all that has been begotten.

When he comes down from the womb to breathe on the day he is born,
you open up his mouth completely, and supply his needs.
When the fledgling in the egg speaks in the shell,
you give him air inside it to sustain him.
When you have granted him his allotted time to break out from the egg,
he comes out from the egg to cry out at his fulfillment,
and he goes upon his legs when he has come forth from it.

How plentiful it is, what you make,
although they are hidden from view,
sole god, without another beside you;
you create the earth as you wish,
when you were by yourself, » before ...
mankind, all cattle and kine,
all beings on land, who fare upon their feet,
and all beings in the air, who fly with their wings.

The lands of Khor and Kush
and the land of Egypt:
you set every man in his place,
you allot their needs,
every one of them according to his diet,
and his lifetime is counted out.

Tongues are separate in speech,
and their characters / as well; their skins are different,
for you differentiate the foreigners.
In the underworld you make a Nile
that you may bring it forth as you wish
to feed the populace,
since you make them for yourself, their utter master,
growing weary on their account, lord of every land.

For them the Aten of the daytime arises,
great in awesomeness.

All distant lands
you make them live,
for you set a Nile in the sky
that it may descend for them
and make waves upon the mountains like the sea
to irrigate the fields in their towns.
How functional are your designs,
Lord of eternity:
a Nile in the sky for the foreigners
and all creatures that go upon their feet,
a Nile coming back from the underworld for Egypt.

Your rays give suck to every field:
when you rise they live,
and they grow for you.
You make the seasons
to bring into being all you make:
the Winter to cool them,
the Heat that you may be felt.
You have made a far-off heaven in which to rise
in order to observe everything you make. Yet you are alone,
rising in your manifestations as the Living Aten:
appearing, glistening, being afar, coming close;
you make millions of transformations of yourself.
Towns, harbors, fields, roadways, waterways:
every eye beholds you upon them,
or you are the Aten of the daytime on the face of the earth.

When you have gone
[no] eye can exist,
for you have created their sight

o that you (yourself) are not seen
(except by your) sole [son] whom you have made.

You are in my heart,
and there is no other who knows you
except for your son (Nefer-kheperu-Re Wa-en-Re) |,
for you have apprised him of your designs and your power.
The earth comes forth into existence by your hand,
and you make it.
When you rise, they live;
when you set, they die.
You are a lifespan in yourself;
one lives by you.
Eyes are / upon your perfection until you set:
all work is put down when you rest in the west.
You who rise and make everything grow
for the King and (for) everyone who hastens on foot,
because you founded the land
and you raised them for your son
who has come forth from your body,
the King of Upper and Lower Egypt, the one Living on Maat,
Lord of the Two Lands (Nefer-kheperu-Re Wa-en-Re),
son of Re, the one Living on Maat, Master of Regalia,
(Akhenaten) |, the long lived,
and the Foremost Wife of the King, whom he loves,
the Mistress of the Two Lands, (Nefer-nefru-Aten Nefertiti),
living and young, forever and ever.

(Simpson and Ritner 279–283)

Truth exists in his Truth.
Pen, papyrus, and palette of Thoth
Avoid the doing of evil.
Beauty is beautiful when beauty is for him
Now truth is for eternity;
It goes down with its doer to the necropolis.
When he is buried, the earth unites with him,
And his name is not smeared on earth.
He is remembered because of goodness;

This is the rule of the word of God.

(Fisher 25)

Bibliography

Erman, Adolf. *Ancient Egyptian Poetry and Prose*. Translated by Aylward M. Blackman. Dover Publications, 1995.

Fisher, Loren R. *The Eloquent Peasant*. Cascade Books, 2015.

Lichtheim, Miriam. *Ancient Egyptian Literature Volume 1: The Old and Middle Kingdoms*. University of California Press, 2006.

Sesanti, Simphiwe. "Ancestor-Reverence as a Basis for Pan-Africanism and the African Renaissance's Quest to Re-Humanize the World: An African Philosophical Engagement." *International Journal of Social Science Studies*, vol. 5, no. 1, January, 2017, pp. 63–72.

Simpson, William K. and Robert K. Ritner. *The Literature of Ancient Egypt: An Anthology of Stories, Instructions, Stelae, Autobiographies, and Poetry*. Yale University Press, 2003.

Afterword
Speaking to Posterity

Maat was the guiding principle of harmony, truth, balance, and order that permeated ancient Kemetic political and civil life. Although evolving in form and practice over the centuries, Maatian ethical principles can be traced all the way to Kemet's prehistory. My goal has been to provide a rigorous and vigorous, communication-focused account of the ethical wisdom (rules of engagement) ancient Kemites cultivated and is evidenced in the form of recovered written texts, stelae, and the Medu Netcher script itself. Throughout the book, I've labored to present Maatian ethical teachings and communication practices as, at least partly, responsible for the longevity and success of Kemetic culture. At stake is the recognition that, in spite of widespread views of ancient Kemetic culture as blood-thirsty conquerors who mostly ruled by brute force, Kemites also promoted ethical philosophies grounded in communitarian wisdom that paid careful and sustained attention to listening habits and listening itself as the foundational moral virtue.

Maatian ethics are grounded on the spiritual dialectical laws of opposites that guided the practice of "medu nefer" (beautiful, good speech). In this context, speech was seen as sacred, constitutive of reality, creative, and transformative. Above all, Maat was an individual and collective means to keep *isfet* (evil, chaos, ugliness) at bay. An ethical approach to communication is mindful and reflective of such communicative ends. In other words, to communicate ethically is not simply to figure out "the just thing to do" but it is concerned with the pursuit of the good, a good that may or may not fit neatly into the domain of justice.

One particular realm of our contemporary society where I feel this tension becomes most salient is in the discourse of diversity. We are living in societies that are increasingly more diverse, globalized, and encompassing of an ever-growing number of collectivities, belief systems, and personal preferences. So, what does it mean to communicate ethically in this sort of society?

In the United States, there are those who think that the fabric of our culture is being ruptured by multiculturalism. These voices need not be anti-semitic or bigoted, but well-meaning cultural critics who worry about our diminished ability to make moral claims on one another. Cultural critics like Morris Berman in his book *The Twilight of American Culture* and Christopher Lasch in *The Revolt of the Elites* voice their concern that the rise of a multicultural ethos in American society has left us unable to make moral demands and claims of each other and this has led to a decline in American democratic debate.

Lasch states, "'Diversity'—a slogan that looks attractive on the face of it—has come to mean the opposite of what it appears to mean. In practice, diversity turns out to legitimize a new dogmatism, in which rival minorities take shelter behind a set of beliefs impervious to rational discussion" (17). He continues, "each group tries to barricade itself behind its own dogmas. We have become a nation of minorities, only their official recognition as such is lacking to complete the process" (17). He concludes: "one begins to slide down the slippery slope to relativism, moral anarchy, and cultural despair" (243).

In the previous passage, it seems clear that rather than taking issue with the idea of multiculturalism itself, Lasch is taking a shot at what he considers to be a poor/detrimental instantiation of it. In advancing an agenda of pluralism, a discourse of rights framed in terms of justice has been provided as a reason to be tolerant and cognizant of a range of perspectives and cultures, thereby effectively banning possible criticisms to be made of others' views. In relying on a discourse of rights (justice), moral/ethical demands on others are made impossible according to Lasch.

So, when it comes to social change and the transformative communication ethic that can bring it about is a discourse of ethics/morality more suitable, more effective than one of justice? Vice versa? Or should we strive for a combination of the two? Surely, the reasonable way to begin to answer this question (like any other good philosophical response) is . . . well . . . it depends.

Hegel's idea of social transformation is principled and political. It also requires its individuals and societies to essentially move through three stages (*aufhabum*): (1) Thesis (negation); (2) antithesis (preservation); and (3) synthesis (superseding). Holding together Hegel's teleology is the idea that moral judgments are never made in the abstract nor are they universalizable beyond context. However, many readers of Hegel come away with the impression that when Hegel talks about universals, he does so as to say that context, particularly, historical context is irrelevant. Not so. Hegel's project (contra Kant) is to find some way to escape having to provide a metaphysical justification for his teleology. This is a tricky proposition but ultimately not an unintelligible one. A moral code for Hegel replaces an external teleology in moving a society forward. This moral code or internal teleology is shaped, discovered by dialogue, rooted in its historical moment and material conditions, and while universal, it is changeable over time. Once again, universal is not meant as atemporal or suprahuman but rather applicable to a particular a collectivity in a particular place in time. Even the *begriff* or the structure by which a society is to engineer its transformation is changeable.

In the *Philosophy of Right*, Hegel states that "education is the art of making man ethical" (350) and Paulo Freire perhaps emphasizes this point more than any other in his own work. By the time Freire writes "Pedagogy of the Oppressed" in 1970, he had studied with great interest the writings of Hegel, with special attention to Hegel's master–slave dialectic as it would later help him formulate his own theory of relations of domination and the dependency that develops between master and slave as part of their mutual recognition as such. Freire also likely found Hegel's teleology of human potential empowering and useful to the marginalized Latin American poor seeking to attain an identity and a consciousness of their own throughout most of the 20th century.

At first glance, the concept of education found in the body of work of Brazilian philosopher of education Paulo Freire and Hegel seems to bear a striking resemblance. For instance, they both speak of education as a vital part of the process of becoming fully human and both Freire and Hegel emphasize the dialectical attributes of a truly liberatory educational practice. Both argue that the proper structure of education can transform societies. Additionally, they believe that the community should act as the educating body that facilitates the actualization of its own general (universal) will. In short, both Hegel and Freire make education a cornerstone of their larger concern for humanization or the full actualization of human potential.

But there are important differences between their projects. For Freire, the central problem is this: How can the oppressed, as divided, unauthentic beings, participate in developing the pedagogy of their liberation? In response, he invests education not just with the power, but the responsibility to challenge social norms and institutions (51). Freire believes in what education can do to help illiterate workers, and other disenfranchised groups, and the ways in which education can be put at the service of social revolutions.

So what do this mean for a transformative communication ethic? The dialectic process, as an ethical stance, is the ultimate ethical commitment in the transformational process of a society. Thus, the merging of old and new and divergent ideas is also an ethical means to promote change. Freire is an example of historically appropriate communicative action, while Hegel was emphatic that we must not privilege subject over substance. A pluralistic society develops an internal teleology, an exo-skeleton that is contextualized and reflective of its components and material conditions and by extension its laws become instantiations of its cultural values.

As Dr. Ron Arnett puts it in his "Paulo Freire's Revolutionary Pedagogy: From a Story-Centered to a Narrative-Centered Communication Ethic": "A philosophy appropriate for a world of diversity must contend with the diverse interests of social classes and shifting demands of the historical moment" (160). "Interaction among individuals, a given story, and the historical moment lessen reification of a tradition while lessening relativism, because the story offers guiding coordinates" (161).

Hegel and Freire made invaluable contributions to social revolution. Now is our turn, yours and mine. The current canonical arrangement of rhetoric, communication studies, and ethics is in need of revision. Maatian ethics must be given their initiative place in the study of communication, rhetoric, and communication ethics. We have an ethical obligation and the opportunity to correct the mistakes and omissions of the past. We must promote intellectual justice, and give credit where credit is due, so truth can prevail. Let the ancestors speak!

Bibliography

Arnett, Ronald C. "Paulo Freire's Revolutionary Pedagogy: From a Story-Centered to a Narrative-Centered Communication Ethic." *Qualitative Inquiry*, vol. 8, no. 4, 2002, p. 489.

Freire, Paulo. *Pedagogy of the Oppressed*. Continuum, 2000.

Hegel, Georg W. F. *Hegel's Philosophy of Right*. Translated by T. M. Knox. Clarendon Press, 1957.

Lasch, Christopher and W. W. Norton & Company. *The Revolt of the Elites: And the Betrayal of Democracy*. W.W. Norton, 1995.

Index

Note: *Italic* page numbers refer to figures

abstraction 55
Ammit 24
Aristotle xii, xiv, 18–19, 74
Arnett, Ronald C. *114*
Asante, Molefi Kete xv, 84
Assmann, Jan 5, 81
Atum ix

Ba 28
Beatty Papyrus IV, 74
Benhabib, Seyla 51, 54–55
Bernal, Martin 55–56, 74, 80, 82
Book of Coming Forth by Daylight
 24, 68, 70
Budge, E. A. Wallis 24

The Cairo Lone Songs 96
Champollion, Jean-Francois 81
Christians, Clifford G. 2, 53
Clarke, John Henrik xiv
cosmology vii, ix, 3, 20, 38, 67
Critique of Judgement 59, 61, 63,
 90–91
Cyena-Ntu (C-N) language family 7

Darwin, Charles viii, 23
The Declarations of Innocence 24–26
Dedication Inscriptions of Seti I 95
Democritus 74
Denzin, Norman K. 4
Descartes, Renee 18, 59
Diodorus Suculus x–xi, xv

Diop, Cheikh Anta 57, 60, 74
Djehuti (Thoth, Tehuti, Hermes)
 viii, ix, 21, 23, 40
doing Maat 27
Duat 17, 28, 70

The Ebers Papyrus 45
The Edwin Smith Papyrus 45

fields of the blessed *28*
42 Negative Confessions 24–26
Freire, Paulo 90, 113–114

Geb 35
The Great Hymn to the Aten 104–109
Green Sahara Periods (GSPs) 3–4

Hathor 40
Hegel, G.W.F. 64, 90, 112–114
heka 7–9
Heraclitus 9
Herodotus xiv, xv
Hu 35
Hypatia 45

Imhotep, Asar 7
*The Instruction of Little Pepi on His
 Way to School* 45
The Instruction of King Meri Kere
 72, 77
Isfet 13, 62, 111
Isocrates xiii, xiv

Ka 24, 28, 65, 73
Kant, Immanuel xiv, 25, 36, 50–65, 87–88, 90
Karenga, Maulana 28–29, 38, 69, 73
Kesh Temple Hymn (Liturgy to Nintud) xii
Kierkegaard, Soren 34–37, 42–43, 54, 64–65
Kush xii, 7

The Lamentations of Khakheperre-Sonb 99
Library of Alexandria *73*
logos 9
The Love Songs of Papyrus Chester Beatty 87

maa kher 57
Maia 40
maieutic artistry 38, 47
The Man Who Was Weary of Life 99
Marx, Karl *58*
MDW NTR (Medu Neter. Mdw Netcher) 5, 7, 13, 29, 40, 62, 68, 74, 77–92
medu nefer 72
Mill, John Stuart xiv, 59

Nefer 63
Neteru 24
NTR 21, 80
Nietzsche, Friedrich 90
Nimrod xii
Nubia xii, 7, 81
Nun 21–22, 35–36
Nut 22

Obega, Théophile 28, 47, 63, 75–76
Osiris 35

The Papyrus Harris 500 96
Plato xi, xiv
Pleione The Oceanid 40
Prayer and a Hymn of General Haremhab 95
Ptah ix, 21, 35, 67
Pythagoras of Samos xiii, 43

Ra (Re) 23, 35
Ra, Ankh Mi 95
Rekhet 63

Seshat 72
School of Alexandria 45–46
Shu 35
Sia 35
Socrates xi, xiv, 29, 38–40, 71
speaking Maat 27, 37
Spell from Passing Akhet to the Sky (Part of the Pyramid Texts of Pepi I) 56
The Stelae of Intef, Son of Senet 74

Ta-Seti 7
The Teachings of Ptahhotep 73, 77, 98, 102
The Tale of the Eloquent Peasant 95, 98, 102, 104
Theory of Correspondence ix
The Trees in the Garden 104
Trismegistus, Hermes 40
Triumphal Hymn of Ascension (Pyramid Text 51.1) 27

Wiredu, Kwesi 42, 64
The Wisdom of Anii Papyrus 76

Zeus 40

Printed in the United States
by Baker & Taylor Publisher Services